CAROLINE GORDON

CAROLINE GORDON

A BIOGRAPHY

Veronica A. Makowsky

New York   Oxford
OXFORD UNIVERSITY PRESS
1989

Oxford University Press

Oxford New York Toronto
Delhi Bombay Calcutta Madras Karachi
Petaling Jaya Singapore Hong Kong Tokyo
Nairobi Dar es Salaam Cape Town
Melbourne Auckland
and associated companies in
Berlin Ibadan

Published by Oxford University Press, Inc.,
200 Madison Avenue, New York, New York 10016

Oxford is a registered trademark of Oxford University Press

Library of Congress Cataloging-in-Publication Data

Makowsky, Veronica A.
    Caroline Gordon : a biography / Veronica A. Makowsky.
        p.   cm.
    Includes index.
    ISBN 0-19-505718-X
    1. Gordon, Caroline, 1895–   —Biography.   2. Novelists,
American—20th century—Biography.   I. Title.
PS3515.O5765Z74 1989
813'.52—dc19
    [B]   88-29301
                                                         CIP

9 8 7 6 5 4 3 2 1

Printed in the United States of America
on acid-free paper

In Memory of Veronica Popylisen

PREFACE

I first heard of Caroline Gordon when I was in graduate school at Princeton. While working on some of R. P. Blackmur's unpublished manuscripts in Special Collections, I was flipping through the manuscripts' catalogue when I noticed that the library had acquired a considerable collection of papers from someone named Caroline Gordon (1895–   ). A librarian was standing nearby, so I idly asked who Caroline Gordon was. "Allen Tate's wife. Oh, and she wrote fiction herself" was the reply.

The Fugitive and Agrarian poet Allen Tate was familiar to me, but I knew nothing about his spouse who also wrote. Ever on the lookout for good fiction, I began to borrow her novels and stories from the library until I had read most of them; then a question inevitably arose: why was a woman with such a body of serious work, nine novels and two collections of short stories, so relatively unknown? As I began to look through her papers, the thought of trying to write a biography tentatively emerged. With the encouragement of Richard Ludwig and Alfred Bush, who had known her when she lived in Princeton, I wrote to Caroline Gordon in Mexico where she lived with her daughter and son-in-law, sending her some of my work and asking if she would consider authorizing a biography. She would and did, and so this work began.

In March 1981, I visited Caroline and her family, and, with their helpful information and encouragement, began a round of travels and interviews that lasted several years. As I acquired copies of her letters from around the country and delved into her papers and published works, my initial query about her relative obscurity became directed toward questions about the place of a woman writer in a literary world dominated by masculine imaginations. What formed Caroline Gordon's imagination and the themes of her work? Who set her standards for serious art? Did male mentorship hurt or benefit her work and her reputation? These are some of questions this biography attempts to answer, but whether it succeeds or fails, I hope that it will lead more readers to Caroline Gordon's work where they can judge it for themselves, and enjoy it.

I would like to thank Nancy Tate Wood and her husband Dr. Percy Wood for their hospitality, encouragement, information, and suggestions. Though the word is overworked, their help was truly invaluable.

For the grants that gave me the time to complete this work, I would like to thank the National Endowment for the Humanities, Louisiana State University, and the Newberry Library.

For interviews I am grateful to Cleanth Brooks, Brainard and Frances Cheney, Jean Detre, Malcolm Cowley, Sally Fitzgerald, Roger Hecht, Mimi Laughlin, Andrew Lytle, Willard Thorp, William Tillson, Leonard Unger, Ray B. West, Jr., and Sally Wood (Mrs. Lawrence Kohn).

For assistance by correspondence, I wish to thank Ursula Beach, Frances Bennett, M. E. Bradford, Ashley Brown, Peter J. Casagrande, William Combs, Helen Conyers, A. C. Edwards, John Ferguson, John Goetz, Cynthia Gooding, Polly Ferguson Gordon, Edward Grier, Allen Jossey-Bass, David Hallman, George M. Herman, John Howett, Marjorie Kaplan, Robert Kettler, Mrs. Charles R. Kirk, Sister Jean Klene, Sister Mary Immaculata, and Sister Bernadette Marie at St. Mary's College; Meridel Le Sueur, Janet Lewis, Catherine Patterson Maccoy, Harvey Lyon, Mrs. James L. Major, Mrs. C. H. Moore, Gordon O'Brien, Mary O'Connor, Katherine Oakley, Danforth Ross, James Ross, Richard Schwab, Kenneth Silverman, Sondra J. Stang, Norma H. Struss, W. J. Stuckey, Mary Barbara Tate, Mrs. John Dargan Watson, Floyd C. Watkins, Harold H. Watts, Manfred Weidhorn, Edgar Wolfe, George J. Worth, and Celeste Turner Wright.

I am grateful to Janice Biala for permission to quote Ford Madox Ford. For access to Caroline Gordon's wonderful letters to him, I am deeply grateful to Ward Dorrance. I wish to thank Kathleen Donahue for access to John Berryman's Papers at the University of Minnesota Library. To Elizabeth Hardwick I am grateful for access to letters in Houghton Library's Lowell Collection. For a striking photograph of Caroline Gordon, I am indebted to Mrs. Gerald Lambert. I wish to thank Robert Liddell Lowe for access to Gordon's letters in his papers at the University of Tulsa Library. To Robert Penn Warren, I am grateful for access to Gordon's letters to him in the Yale University Library.

The Princeton Library is the repository of Caroline Gordon and Allen Tate's Papers as well as the Scribner files, and to the staff of that library I owe a considerable debt of gratitude for innumerable acts of assistance over the past seven years. With the assistance of the English Department, Library, and Registrar's Office of Bethany College, I was able to find much information about Caroline Gordon's college years. The Department of English and the Library at the University of Minnesota were most helpful about the Tates' Minnesota years. I am grateful to Robert Giroux for access to Gordon's author file at Farrar Straus Giroux. The Interlibrary Loan Department at Middlebury College was also of great assistance. I am grateful to the Macmillan Publishing Company for use of the Scribner Collection at the Princeton University Library.

In addition, I would like to express my gratitude to the following institutions: the American Academy and Institute of Arts and Letters, Amherst College Library, University of Arkansas Library, Boston University's Mugar Memorial Library, Brown University Library, Mark Van Doren Papers in the Rare Book and Manuscript Library of Columbia University Library, University of California at Berkeley's Bancroft Library, Emory University's Woodruff Library, University of Florida Library, Georgia College's Russell Library, Georgia Historical Society, University of Georgia Library, Harvard University's Houghton Library, Kentucky Library at Western Kentucky University, Kenyon College's Chalmers Memorial Library, University of Illinois Library at Urbana-Champaign, Indiana University's Lilly Library, Newberry Library, Memphis State University Library, Middlebury College Library, University of Minnesota Library, the Southern Historical Collection of the University of North Carolina Library at Chapel Hill, University of Notre Dame Archives, University of Oklahoma Library at Norman, State Library of Pennsylvania, Rice University's Fondren Library, Rhodes College's Burrow Library, Southern Illinois University's Morris Library, Department of Special Collections and University Archives of Stanford University Libraries, Syracuse University's Arents Research Library, University of Texas at Austin's Harry Ransom Humanities Research Center, Special Collections in the McFarlin Library at the University of Tulsa, Utah State University's Merrill Library, Manuscripts Division of the Special Collections at the University of Virginia, University of Washington Library at St. Louis, University of Washington Libraries at Seattle, Washington and Lee University Library, Western Reserve Historical Society, Wilmington College of Ohio's Archives, and the Collection of American Literature at Yale University Library's Beinecke Rare Book and Manuscript Library.

And, once again, for the steady support of my husband, Jeffrey C. Gross.

Baton Rouge   V.A.M.
October 1988

CAROLINE GORDON

CHAPTER 1

T he first thing I remember in this life is my attempt to take that life, declares Caroline Gordon at the beginning of her memoirs. One late afternoon in her grandmother's chamber in Kentucky, the four-year-old Caroline was left alone for what seemed quite a long time. She found herself mysteriously drawn to the basin of water near the window.

> The shadows deepened. With infinite caution I shifted my gaze to where the water glinted. It seemed to me that the shadows swayed forward. I gave way to panic and ran across the room and thrust my face down into the water in the basin. It seemed to me deep enough to drown in. One of the sharpest memories of my life is the surprise I felt when my childish visage raised itself, apparently of its own accord, and I knew that I was still there in that room, with only the shadows for companions.[1]

The sense of abandonment to menacing presences, the moment of panic and despair, the seemingly miraculous recovery, and the resolution to confront the danger once more—these elements constitute the continual scenario of Caroline Gordon's life, the pattern that made her the good artist she was and prevented her from becoming the great artist she might have been. "I, myself, cannot remember when I was not aware that life was a desperate affair at best," she wrote of her childhood. To account for this state, the small girl characteristically invented a story, a cosmic version of her sense of abandonment in her grandmother's room: the world "had been created as a plaything by a group of men, who, tired of sporting with it, had gone on to other pleasures, leaving it to roll on the way it would."

She is not using "men" in its generic sense, for to Caroline Gordon the male's ability to create, to sport, and to perform heroic deeds would always be counteracted by his tendency to desert it all at whim and leave women to suffer the consequences. Since men are the creators in this myth, the women artist is an anomaly; she is hubristic, vulnerable, and necessarily second-rate. "Women," Gordon wrote, "unless their deepest instincts are perverted by false education (as happens so often nowadays), are immemorially inclined to take life itself for granted. For them the mortal condition is the climate in which they naturally move. It is the wayward, tender, flickering masculine intellect which, playing torch-like on the verge of the abyss, even while it illuminates its depths, can easily conceive of life being hurled backwards, downwards into that void of non-being." With this statement Gordon rendered herself a paradox or an impossibility, for she herself was certainly aware of the abyss and flirted with it in her life and art.

The source of this fundamental contradiction in Gordon's sense of self is found in her interpretation of her ancestry, both paternal and maternal. She saw in her family background the repeated struggles of men and women, with each other, as well as with life, "that desperate affair." Their willingness to persevere despite the odds caused her to believe that her ancestors were "cast in heroic mould." Since the outcome of life's battle is, by definition, fatal, heroism meant the ability to confront certain defeat. What she perceived as the inevitable limits to her family's heroism were those she also believed restricted her own accomplishment, in what often proved a self-fulfilling prophecy. Her claim that "the story of their lives . . . is the story of my own" is true; she made it her own in her life and fiction, and, to some extent, she also made the story.

Caroline Gordon's version of her paternal ancestry is consistent with her characterization of men in her life and works. According to her, the Scottish Gordons

> were one of the most ubiquitous of all the clans and on record as the most treacherous. . . . The Adam of Gordon who appears in the ballad has always seemed to me a proper 'father image' for the entire clan.

The ballad records that as the weather turned cold, he remarked to his followers that it was time to seek winter quarters. He and his followers therefore 'drew up' to the nearest 'hold.' The lord of the manor was away from home. His lady appeared at the window and in her frenzy yielded to Adam's smiling promise that he would take care of any bairn she threw down to him. The child she threw down was a girl child of two or three, so rosy and plump that Adam, as he spitted her on his sword, observed that this was the first child he had ever regretted treating in this fashion.

His descendants continued in the ways of this most unpropitious father image down the path of destruction by helping Cromwell loot Irish churches, so that, Gordon wrote, "like many traitors in those days, they were rewarded for their services to their country by the gift of an Abbey." In Caroline Gordon's version, the new Scotch-Irish Gordons next proceeded to dissipate their new wealth with orgies of hunting and fishing, so that they needed to emigrate to America to recoup their fortunes. To Gordon, the characteristics of her paternal ancestors, her own "father image," include betrayal, treachery, cruelty to women, and dissipation through sport, traits with which she also freely endows many male characters in her fiction.

Armistead C. Gordon, another family chronicler, provides a different account of the family's Irish history, one that Caroline Gordon must have known since she owned a copy of his book. According to Armistead C. Gordon, the Gordons did not earn their abbey at Newry by devotion to pillage, but through the religious zeal of "the Reverend James Gordon, of Cumber, also a town in County Down. This reverend gentleman was a Scotch chaplain in the regiment of Lord Montgomery, a constituent part of Cromwell's invading army."[2] This exemplar of the church militant left the lands to his son James of Sheepbridge, who, as Armistead C. Gordon emphasizes, had a remarkable mother-in-law, Jane Wallace. A mother image unmentioned by Caroline Gordon, this brave widow restored her family's fortunes following the burning of the town of Newry by displaying "a strength of mind superior to difficulties." Armistead C. Gordon also stresses the thrift and prosperity of the Scotch-Irish, but does conclude that the family property "which continued to remain in the possession of

the Gordon family, though with steady diminutions from generation to generation, due to the hospitality, the free-living and the
sporting proclivities of its successive owners, until the mansion
house and a remnant of something more than one hundred acres
were left, was sold in 1902."

Early in the eighteenth century, two grandsons of James Gordon
of Sheepbridge, James and John, emigrated to Virginia where they
continued to display the family's conflicting inclination toward accomplishment and piety on the one hand and sport and dissipation
on the other. Both brothers were successful tobacco farmers, but
John fell into debt through fighting and gambling. James was both
prosperous and pious, but added to his wealth through slaves, a
distillery, and a lottery. One might expect that the brothers' contra-
dictory tendencies would be reinforced by the marriage of their
children, Elizabeth, daughter of James, and James, son of John,
but this younger James began the family's climb to achievement by
serving as a delegate first to the Virginia Assembly and then to the
Virginia convention of 1788 which ratified the federal constitution.

James was also the father of the family's preeminent success,
Caroline Gordon's great-grandfather William Fitzhugh Gordon.
He was a general in the War of 1812 and served as a delegate to the
Constitutional Convention of 1829–30. He is best known for his
career in Congress where he fought what he viewed as the excesses
of Jacksonian democracy, an aristocratic proclivity that Caroline
Gordon emphasized in her notes on the subject.

If William Fitzhugh Gordon expected to found his own noble
dynasty, his hopes were disappointed by his son and namesake.
The younger William Fitzhugh Gordon, Caroline Gordon's grandfather, early revealed his character in a letter he wrote to his father
who was in Washington for a Congressional session: "Dear Father,
This is my first letter and you must excuse my blunders. We have
had very cold weather and brother Ruben and myself have not
gone to school for a week. Mamma tells me to write you how I
have passed the time. I cannot tell exactly, sometimes hauling
wood, sometimes playing with [indecipherable] a little and sometimes reading from books of Homer's Iliad. I like old Homer very
well and just begin to get acquainted with the Gods and God-
desses. I must conclude my letter." His need to seek excuses for his
deficiencies, his love of the classics, and his lack of interest in

public concerns remained characteristic of the second William Fitz-hugh Gordon throughout his life.

He did at least begin to emulate his illustrious father by becom-ing a lawyer, but was spared the practice of law and any further attempts to follow in his father's footsteps when he married Nancy Morris, a heiress with a six-hundred acre estate, Oakleigh. He withdrew to the country so completely that he even failed to visit his mother in her last years. To one of his cousins, he sent letters that are masterpieces of rationalization. After missing his mother's funeral, he wrote to his cousin Mason, "It is a source of sorrow to me that I saw so little of her in her declining years, but she knew that no son ever loved and admired a mother more than I did."

In his rural seclusion, William Fitzhugh Gordon devoted himself to reading the classics and writing a blank-verse version of Sir Walter Scott's *The Bride of Lammermoor*. While these pursuits may have been laudatory, they did not feed his numerous progeny. One daughter remembered him at the head of the dinner table, where he recited "his own poetry—or Shakespeare's—and we chil-dren watched to see if the preserves dish went around." His liter-ary interests were once more coupled with the consumption of food when his library perished in a smokehouse along with the hams.

William Fitzhugh Gordon did not seek all his inspiration from the Muse; he also drank, which helped the family fortunes as little as his poetry did. Late in life, when he was asked why he had given up drink, he replied, "Poverty, by God, has come to my relief!," displaying the sardonic humor he passed down to his son James, born on February 7, 1861, and his grandaughter Caroline. His wife, Nancy Morris, early quitted this world and her unworldly husband, leaving the cares of housekeeping to her oldest daughter Patty.

Whatever his other failings, William Fitzhugh Gordon provided his children with an excellent classical education. Caroline Gor-don's father, James Maury Morris Gordon, remembered the scorn in his father's voice as he announced to his erring heir, "That, sir, is the dative!". He later sent young James to study with his Aunt Vic, an excellent classical scholar and a devout Catholic, perhaps a model for Caroline Gordon in her own love of the classics and conversion to Catholicism. James Gordon received his last formal education at the University of Virginia while living in the "office," a small outbuilding at his Uncle Mason's house in Charlottesville. He did not graduate, but left under mysterious circumstances.

James Maury Morris Gordon is the prototype for the hero of Caroline Gordon's second novel, *Aleck Maury, Sportsman* (1934), and like Aleck Maury, he remembered his childhood as bleak and impoverished, but illuminated by two consolations—his growing devotion to the classics and his love of sport. Unfortunately, when his university career somehow failed, he could not return to his beloved hunting grounds in Virginia because his father's worn-out tobacco lands could not provide him with a living. In 1883 he was compelled to turn his back on the classics and sport and seek his fortune in the West. Since his older brother Willy had found employment as a civil engineer, James Gordon also tried that line of work, first in Kansas City, then in Seattle. A letter to his sister Patty, written from Seattle, demonstrates his characteristic concerns.

> Your welcome letter was indeed a surprise to me, for I had almsot begun to think I was never going to see your handwriting again. I read it last Sunday, just a week ago from today, just as I was leaving town on a survey trip, and could not answer it until I got back, for we were veritably in the woods, sleeping in a log cabin and doing our own cooking. You would have wondered to see me, who used to hate to get out of bed at 9 o'clock, crawling out at half past five, going to the spring for water and slicing up fat bacon to fry for breakfast. I never enjoyed meals more in my life, and we got along all right save for a mishap with the beans. I had begun to pride myself, and brag not a little on my cooking, so one evening the chief of the party sent me into camp early to put on some beans to have for supper, telling me at the same time to boil enough to have some left to bake for breakfast. It was my first trial at cooking beans, and I knew they would swell, but I thought the boys would all be hungry and I had better have too many than too few. So I put about two quarts into a small sized pot and set it to boiling. Well, I can't describe the time I had. I was taking that pot off the fire every five minutes and taking out beans, till I had every *dish, plate,* and *cup* in the cabin *full,* and just as the party came in I was commencing to fill the *water bucket* with beans!
>
> . . . I long to be back at home so often, and think now Virginia must be the loveliest climate in the world. I am so tired of living with people with whom I have nothing in common. Everything here is *Northern* in sentiment, and most of the people are from the state of Maine or

Vermont, which two states I hate worst than any others in the Union. And yet I have made some warm friends here, such as I believe would stick to me through any emergency. They have somewhat of a queer fashion on this coast. The young men are all the most confirmed wanderers I ever saw and each one has a 'pardner.' When two fellows are 'pardners,' everything is in common—they share a common purse, travel together, and one will not take a job in a town unless the other can get one too. I have noticed though, that this sort of thing is very onesided, and in most such alliances the one member keeps the other up. Consequently, I have no 'pardner'. . . .

This letter displays not only James Gordon's youthful high spirits, but his lifelong traits. He loved to eat; his in-laws would later remark resignedly, "All Gordons eat like pigs." His passion for the outdoors would continue to call him from his characteristic indolence. Also enduring was his gift of satire which he directed at himself as well as those despised Yankees. This letter most importantly demonstrates his essentially solitary nature. Life's pleasures and vicissitudes were best experienced alone. The help of a "pardner" was not worth the potential burden of 'keeping the other up." His sense that "alliances" are necessarily entangling presages the difficulties of marriage for someone who would himself become a "confirmed wanderer."

James Gordon's bachelor wanderings were soon ended when a bout of typhoid fever caused him to return to Virginia for a protracted convalescence. Since opportunities for earning a living there had not improved and his health precluded more adventures in the West, his family sought a job for him as a tutor in the extensive Meriwether "connection" in Kentucky, into which a cousin had earlier married. He boarded with Douglas and Caroline Meriwether of Merry Mont farm, near Trenton, whose children were among his students. One daughter, Nancy, a promising student of the classics, soon became his bride, and James Gordon and his seventeen-year-old wife settled down to life among the Meriwethers.

As Caroline Gordon wrote, "pride and provincialism are characteristic of most of the Meriwethers." She shared the pride, but missed the provincialism since she was born a wandering Gordon and married the roving Allen Tate. She did, however, identify herself with the Meriwethers in many ways and seems to have acquired some of their principal traits. Louisa Minor, the family historian, summarizes these characteristics in *The Meriwethers and Their Connections*.[3] On the positive side, the Meriwethers were "a big-hearted race—hospitable in their homes to a fault." They were particularly notable for their "genuine love for their kinsfolk,— and claim relationship to the hundredth generation." Caroline Gordon was always famous for her prodigious hospitality, and included invalid or indigent relations in her largess. Throughout her travels, she remained in touch with cousins of many degrees and so never lost her Meriwether connection.

The Meriwethers' family loyalty, however, became somewhat insular after several generations in Kentucky. They mainly stayed in those homes from which they were dispensing hospitality and intermarried within their family and neighborhood. Some family virtues become inbred into faults. Louisa Minor writes, "The Meriwether characteristic from the first Nicholas has been firmness, but of those of this latter day, many people say it has degenerated into obstinacy. . . . An old cousin, who has a full share of the Meriwether blood, and is consequently an authority on the subject, says they are clanish [*sic*] but full of prejudice, and so sincere that it amounts to rudeness." Caroline Gordon, having herself a "full share of the Meriwether blood," is remembered for her tenaciously held opinions and for the steadfastness with which she defended them, a forthrightness sometimes perceived as rude. Louisa Minor concludes of the Meriwethers that "certainly they are 'a peculiar people,' but their peculiarities make them more original, and whatever weaknesses they may develop they ever lean to virtue's side," a summary that could also apply to Caroline Gordon, original and often larger than life in her virtues as well as her faults.

The traits Caroline Gordon wanted to see in her "peculiar people" are found in her description of the family's heraldic device. "The Meriwether arms," writes Gordon, "granted by Richard the Second of the Battle of Bosworth, to Henry Aylworth Meriwether show forth the family's rustic origin. A sun smiles on an azure field for 'Merry Weather.' Sheafs [*sic*] of wheat are the first tangible evidence one finds for the Meriwethers' passion for owning land. Bees doubtless connote yeomen industry and the 'mailed hand embowed' and twined with the ubiquitous serpent the 'wisdom in

battle' which got Henry Aylworth knighted on Bosworth Field."
Gordon's reading of the family coat of arms makes them prototypi-
cal Agrarians, devoted to the land and willing to defend it in battle.

Little is known, however, about Gordon's branch of the Meriwethers, so her imagination could play on the few facts and create
a sort of fairy tale. Her branch of the Meriwethers descended from
Welshmen "of low stature, dark complexioned, round headed."
One Nicholas Meriwether, called "the Welshman" in family legend, sent to Virginia three of his gnomic sons, the requisite fairy-
tale number. There they established themselves on land which,
tradition tells us, their father had received between 1652 and 1654
"in return for money which he had furnished Charles the Second
when he was attempting to establish himself on the throne."

In the best fairy-tale tradition, one of the three sons, Nicholas II,
Caroline Gordon's ancestor, became a planter-prince. He fought
battles, won the hand of a fair lady, and founded a noble house,
but only after serving an apprenticeship to a wise and difficult old
man. As a youngster in Virginia, Nicholas was sent to school to
William Douglas, later known as Parson Douglas, a figure who
fascinated Caroline Gordon for his learning, his conservatism,
and his worldly religion, all traits of her own father, James Gordon. Parson Douglas was born in 1708 in Scotland where he was
also educated and began to acquire his famous library. Curiously
enough, while at the university in Edinburgh, he was tutor to the
son of a Mr. James Gordon of Orange. He came to America as a
tutor to James Monroe, later President. At one time, he also
taught both Nicholas Meriwether and Thomas Jefferson. Family
legend has it that in front of his future wife, the parson's daughter
Margaret, young Nicholas often heard the exacting scholar com-
plain, "Nicky Meriwether, why *can't* you be clever like Tommy
Jefferson?"

If poor Nicholas was humiliated in the schoolroom by such a
peerless rival, he won distinction on the field of battle during the
French and Indian War and became known as a "man fearless and
self-possessed in trying places. He was one of the colonels of the
Virginia regiment attached to Braddock's army, who bore the
wounded general from the field of battle after his defeat" near
Pittsburgh on July 9, 1755. In her unpublished memoirs, Caroline
Gordon writes that he was killed performing this act of heroism,

but although her version makes a good story, the truth is necessary to make Caroline Gordon. He survived the battle, and since his heart was not faint, in 1759 he did win the fair lady, Margaret Douglas. Their son, Dr. Charles Meriwether, was Caroline Gordon's great-great-great grandfather.

Margaret Douglas Meriwether was more than a necessary dynastic link. She set the pattern for the strong Meriwether women, like Caroline Gordon, who could outdo their men in loyal, or foolhardy, devotion to a cause. According to Louisa Minor, Peggy Douglas Meriwether and her mother "had strong Tory proclivities during the war which severed us from England and when the husband and father [Parson Douglas] was compelled to swear allegiance to the new Republic of America, they were highly indignant and gave him their opinion of the act in strong terms." Their gift for the rhetoric of moral indignation was also inherited by Caroline Gordon and other women of her family.

Another family connection of great importance to Caroline Gordon was formed in eighteenth-century Virginia. The explorer Meriwether Lewis, protégé of Jefferson and leader of the Lewis and Clark expedition, was born in Virginia in 1774. As his name indicates, his mother was a Meriwether. Caroline Gordon's last unfinished novel, *Joy of the Mountains,* concerns Meriwether Lewis' relationship with Jefferson, his famous explorations, and his mysterious death at a lonely Tennessee tavern in 1809.[4]

Although the novel is ostensibly about Meriwether Lewis, the character of Thomas Jefferson often predominates in the same way Satan sometimes seems the hero of *Paradise Lost.* Gordon's Jefferson is a Meriwether family villain for three reasons illustrative of the family values. Jefferson was opposed to the entail of land, a practice which bound the land so that heirs could not sell it or give it away. This struck a blow against the Meriwether's obsession with gaining land and keeping it in the family. Second, Jefferson's democratic tendencies were considered treachery against the conservative class-consciousness the Meriwethers asserted they nobly exemplified. Finally, Caroline Gordon believed that Jefferson's published account of Meriwether Lewis's death as a suicide from melancholia, not a murder, was a betrayal of the pupil by his mentor, a belief that also betrays Gordon's own patterns of ambivalence toward mentors.

Gordon's fictive treatment of Meriwether Lewis dates from the last decade of her life; the first ancestor she used in her fiction was her great-great-great grandfather, Dr. Charles Meriwether, son of the gallant Nicholas and Margaret Douglas, and founder of the Kentucky-Tennessee connection. Borrowing Henry James's phrase about characters in his imagination, Gordon wrote that Dr. Charles "has solicited me all my life." In her fiction, however, he appears only indirectly: in the time-present of her first novel, *Penhally* (1931), he is already dead, but his wives and the relationships among his sons provide Gordon's inspiration for the strife-torn, fratricidal Llewellyn family. Gordon could not use Dr. Charles himself as a character because he seemed unrealistically perfect. She wrote, "To me the most mysterious thing about Charles . . . is his—impeccability. I have never heard a word said against him and I have never found—in print—any intimation that he had any faults."

Although she considered the impeccable Dr. Charles unsuitable for fiction, she remained fascinated by him. In *Penhally* she evades the question of how such a perfect progenitor created such a conflict-ridden race, really the question of the origin of evil or original sin, but she could not put it aside in her unfinished memoirs, as many fragments and passages show. Her attraction to him is also partially explained by the fact that Dr. Charles, born in 1766, was one of the last nonprovincial Meriwethers before Caroline Gordon herself, who was born nearly a century and a half later. Dr. Charles had the advantage of being the "pet and pride" of his scholarly, worldly grandfather, Parson Douglas, as Caroline Gordon was the pet of her canny maternal grandmother and the pride of her learned father.

Perhaps Charles would have been just another provincial Virginian had he not travelled to his grandfather's native land in 1789 to complete his education. Although he studied medicine in Edinburgh, his interests went beyond its confines. As his letters show, he was a keen observer of agricultural practices in Scotland and revolutionary movements on the continent. He also had an eye for the ladies and discovered a relation of his maternal grandmother who became his first wife. He describes the mellifluously named Lydia Laurie as "pretty of face and shape" and having "every accomplishment of body and mind that can add to his happiness."

He settled in Edinburgh, perhaps to be near his wife's family or perhaps to enjoy Edinburgh's more stimulating atmosphere, but his happiness was short lived. After the death of his wife and infant daughter, he returned to Virginia to begin again at the beginning of the nineteenth century.

In *The Meriwethers and Their Connections,* Louisa Minor charmingly relates the next episode of Dr. Charles Meriwether's romántic career. "In Virginia he was naturally a great toast, being young, handsome, accomplished and wealthy, and was much sought after by the managing mothers, who had marriageable daughters, but he passed by all these snares only to fall a victim to the charms of a little maiden just from school, Nancy Minor . . . who was so hoydenish, she had to be called down from a cherry tree by her 'black mammy,' to receive the addresses of her stately suitor." One might wonder if she should have stayed in the tree since she also died young, but if she had not listened to her "stately suitor," she would not have borne her only child, Charles Nicholas Minor Meriwether, Caroline Gordon's great-great grandfather and the prototype for Ralph Llewellyn of *Penhally* and Fontaine Allard of *None Shall Look Back.*

Perhaps discouraged by his experience with young virgins, Dr. Charles next married a widow, Mrs. Mary Walton Daniel, who appears in *Penhally* as the senile old lady sitting in the window, but who had considerable spunk and acumen in her younger days. With her and his son Charles, Dr. Charles travelled in 1811 to a tract of fertile tobacco land he had purchased "in a green valley between two springs" on the Kentucky-Tennessee border near Clarksville, Tennessee. There he built his house, Meriville, which became "synonymous with large-hearted hospitality," and there he established the reputation for "impeccability" that so bemused Caroline Gordon. In the words of the family chronicler, Louisa Minor, "There could be no finer specimen of manhood than he was. Beloved for his generosity, revered for his wisdom, respected for his unswerving integrity, the very soul of honor, it is impossible to estimate the influence of such a man in forming the morals of the community."

Family legend relates that on the way to Kentucky, Dr. Charles met a young man named Chiles Barker at the Old Graysville Inn. They became such great friends that Barker accompanied him to

Kentucky where he purchased an adjoining tract. Like royalty, they and their descendants enjoyed pointing out how all the land on one side of the Trenton road was Barker land and all on the other side belonged to the Meriwethers. Also like royalty, they affirmed their alliance through the marriage of their children, Charles Nicholas Meriwether and Caroline Barker. Dr. Charles further guaranteed the continuance of his name through two sons by his third wife, William Douglas Meriwether, the model for Nicholas Llewellyn of *Penhally,* and James Hunter Meriwether, the original of the eccentric Jeems of the same novel.

Despite, or perhaps because of, Dr. Charles Meriwether's status as a paragon, his sons did not found the cooperative dynasty he seems to have envisioned. Indeed, there may have been some tension between the half-brothers. Charles Nicholas Meriwether, the eldest son, did not inherit Meriville, but moved to another part of his father's land. Nicholas, the second son, did inherit the home place, perhaps because his mother was still living there. The third son, James, left for Arkansas, only to return to Meriville with his wife and children after his property in Arkansas had been flooded. Caroline Gordon follows the facts of this situation quite closely in *Penhally,* so one can at least speculate that the lack of brotherly love she depicts may also have had a basis in family history.

The three sons did not equal the accomplishments of their father, except in the sense that together their disparate traits did seem to compose a distorted version of their respected sire. The third son, James, inherited his father's scientific bent. He collected dead animals from the neighborhood to make fertilizer from them, an activity regarded as bizarre as well as unsavory. He did not have his father's ability to prosper while pursuing science, so he was tolerated as the abstracted neighborhood eccentric, like Jeems Llewellyn of *Penhally,* who lived on the generosity of his brother Will.

Perhaps William Douglas Meriwether did not mind supporting his brother's family since, as a bachelor, he had no heirs but had accumulated a considerable fortune. His unmarried state, however, did not mean that he had not acquired his father's ways as a ladies' man, but only that he directed them somewhat differently. Of "Curious Will," Caroline Gordon wrote, "his mulatto butler so closely resembled him that one of the ladies of the connection remarked on the resemblance." Caroline Gordon endows his fictive version, Nicholas Llewellyn, with the same proclivities in the opening scene of *Penhally*.

Charles Nicholas Meriwether, Caroline Gordon's great-great grandfather, inherited his father's love of gracious living. He built a large brick house, Woodstock, with a ballroom, where he practiced a prodigious hospitality of the kind we now consider Southern. Family letters indicate an almost constant stream of relations and friends who visited for a night or a year. Woodstock also became a place of refuge for indigent or invalid relations. Charles Meriwether's particular pride was the breeding of champion race horses, often named from the novels of Sir Walter Scott. "The 'Woodstock' stud was famous for four-mile horses that were unbeaten on the southern track," according to Louisa Minor.

Caroline Gordon must have found her great-great grandfather's lavish life a paradigm of that of the southern gentleman because she dedicated her first novel to his memory. *Penhally* is in some ways an Agrarian paean to the lost glories of the South, although Gordon also illustrates the weaknesses inherent in that way of life. Charles Nicholas Meriwether represents the vulnerability as well as the glory. He attempted to maintain his way of life into the Civil War. He hosted lavish barbecues and dances for the soldiers. He donated his horses to the Confederacy, including one ridden by General Johnson at Shiloh. He also gave his eldest son, Ned, who was killed in the Battle of Sacramento while serving under Forrest. His body was laid out in the newly completed ballroom at Woodstock, as was Ned Allard's in *None Shall Look Back* (1937).

Charles Nicholas Meriwether's way of life had been destroyed by the war and with it some of his spirit. Fortunately, his wife Caroline Barker Meriwether was a woman of many resources, the first of a series of strong Carolines in the family. She travelled throughout the region nursing the sick in the family connection. She acted as a clearing-house for family news through her letters. With her customary practicality and resolve, she also faced the problem of the former slaves returning to a now-impoverished Woodstock after running off to join the Yankees at Clarksville. She writes to her granddaughter, "We are still encumbered with women and children, have only two men on the place. We sent Mary and her family to town. Her husband and brother had gone a year ago. Both the Bens left us large families to support, but we will send them off if they do not go."

Caroline Barker Meriwether also faced the problem of two more strong-willed Carolines. The first was her daughter, Caroline Douglas Meriwether Goodlett. According to Caroline Gordon, she was the "first example in our neighborhood of 'the new woman.' " During the Civil War, Cal set up a nursing station in the same ballroom at Woodstock that had recently seen the bier of her brother Captain Ned. Her first marriage proved unhappy; her husband had an unfortunate tendency to leap out from behind his bedroom door, naked, and embrace the maid who was bringing the hot water. Understandably, Cal flew in the face of convention and obtained a divorce. She next married a Confederate veteran, Colonel Goodlett, and began her career of good works in Nashville, where she was responsible for the introduction of drinking fountains for mules in that city. She also helped end the flogging of female prisoners in the state penitentiaries. Her most notable accomplishment, however, was the founding of the United Daughters of the Confederacy in 1894. As a child, Caroline Gordon considered Aunt Cal a "figure of fun," but later regretted not talking more with her. Caroline Goodlett is the model for the betrayed but enduring Aunt Cal in *Penhally* and *None Shall Look Back*.[5]

The second formidable Caroline with whom Caroline Barker Meriwether needed to contend was her granddaughter, Caroline Champlain Ferguson, the grandmother of Caroline Gordon. Caroline Meriwether's daughter, Nancy Minor, had married into a local family. Nancy's husband, John Dickens Ferguson, was a " 'circuit rider' " who "was accustomed to rehearse his sermons aloud as he rode about the countryside," as does the ancient relative who greets Aleck Maury at the train station in *Aleck Maury, Sportsman*. An abstracted divine, he is remembered for having once asked his bemused overseer if "the maize had a uniform aspect." John and Nancy Ferguson led a wandering existence as he followed the call of his ministry as far away as New Orleans. Their life was further complicated by Nancy's chronic ill health. Her letters indicate that she might have suffered from tuberculosis since she complains of chest pains and coughing, but her poor health may have been either caused or aggravated by the constant homesickness for

Kentucky that fills her letters. Because of her health and frequent moves, her daughter Caroline was frequently sent home to Wood-stock to her grandmother.

Born in 1848, young Caroline Ferguson had quite a difficult youth, but she could give as good as she got. After her mother's death she lived with her circuit-riding father and her stepmother. In a letter from her stepmother to her, one gets the impression that the fairy tale had been reversed, and the stepmother was abused by the wicked stepdaughter. Her stepmother attempts to defend herself against the girl's forthright or rude comments, including an accusa-tion that her stepmother was raising her children to marry for money. Caroline Ferguson's tendency to overcompensate for trou-bles and uncertainties by manipulation and an aggressive tongue would remain with her until her death in her nineties.

Her adolescence was interrupted by the Civil War, but its after-math gave her a chance to display her strength. She married a cousin from the Ballardville, Kentucky, area, Douglas Meriwether, who was still recovering from service under Forrest at Chickamauga, Shiloh, and Donelson. Douglas Meriwether is the model for John Llewellyn in *Penhally* and Rives Allard in *None Shall Look Back*. Like John Llewellyn, his spirit appeared to have been exhausted by his experiences in the War. Like Rives Allard, Douglas was from a "kinky-headed branch" of his family which tended to go to extremes in pursuing religious or social ideas. One "kinky-head" freed all his slaves, sent them to Liberia, and unsuccessfully attempted to raise silkworms instead of tobacco. Douglas Meriwether's mother, Su-san, is the model for Susan Allard in *None Shall Look Back* who would give away everything in the house to the needy. She also set up field hospitals during the Civil War. In fact, Susan Meriwether spent so much time on horseback performing her charitable errands that she was known as "Mammy Horse."

Unfortunately, the equally formidable Caroline Ferguson Meri-wether was from the "Anyhow" branch of the family. The "Any-hows" loved gracious living and lavish display and considered much attention to ideas and principles evidence of "kinks" in the brain. It is not surprising that the young couple only spent about a year with Susan Meriwether before returning to the bride's connection. Caro-line Meriwether had inherited a farm, Merry Mont, from her mother Nancy Minor Ferguson, a devoted reader of Nathaniel Hawthorne. The young couple decided to settle there and confront the difficult task of reviving the neglected farm during the harsh years of Reconstruction. After living in a double-pen log cabin, the Old Place, they built a new house known as Merry Mont or Merimont. They eventually prospered and three daughters and a son were born to them. Their second daughter, Nancy Minor Meriwether, married James Gordon, and the mingling of Gordons and Meriwethers produced the most formidable, intelligent, and successful Caroline of them all, their only daughter, Caroline Gordon.

CHAPTER 2

To Caroline Gordon, Merimont and its environs were "associated with a golden light." She called it "a world which was so self-contained, yet so fully peopled and so firmly rooted in time and space that . . . when its name is pronounced I feel a stirring in my heart which no other name can evoke."[1] In the early 1900s, the Meriwethers and the Barkers did seem to form a little cosmos of their own along the Kentucky-Tennessee border. The Meriwether farms were Meriville, Merimont, Eupedon, Fairfields, and Woodstock. The Barkers held sway at Cloverlands, West End, Glenburny, Oaklands, Summertrees, and The Mill.[2] The two families had intermarried with a third, the clerical Fergusons, but local opinion held that each family still exhibited its ruling passion. The Meriwethers loved land, the Barkers wanted money, and the Fergusons were obsessed with tneology.

In her unfinished memoir, Caroline Gordon populates this self-sufficient universe with elder relations who were distinctive, interesting, often original, and above all, nurturing to an observant and sensitive little girl. Caroline looked like a Gordon, instead of a Meriwether, and so her brothers were considered better looking. A photograph of Caroline as a child shows a dark-haired girl with a sad expression and unusually luminous, shining eyes. A photographer, who was to make portraits of all three Gordon children, had thoughtlessly remarked that it was a pity the boys were better looking than the girl, and so in that photograph, Caroline's eyes are shining with unshed tears.[3] Her grandfather Meriwether, a man noted for his kindness and gentle nature, recognized Caroline's insecurity about her looks. "It was from him that I received my first compliment on my appearance. My mother had dressed me in a fresh muslin dress (white, candy-striped with pink) and had curled my lank brown hair—and had sent me down to tip-toe through the front hall, but not so noiselessly that my grandfather did not raise his eyes from his book and tell me that I looked pretty."

Caroline was her grandmother's namesake and her favorite grandchild. She later wrote that "it was my grandmother, who, perhaps, first-recognized those attributes which go to make up the part of me which has remained more or less unchanged during my whole life." Like Caroline Gordon in later years, Miss Carrie was a formidable, sharp-tongued matriarch whose stubbornly held beliefs had a certain eccentric charm. In particular, Miss Carrie abhorred "these modern inconveniences." She is still remembered for her refusal to have screens put in the windows because she did not want to breathe "sifted air." With her unassailable family pride, Miss Carrie denied that Merimont milk tasted of onions since Merimont cows were too smart to eat onions.[4] For Caroline Gordon, Miss Carrie's strength was associated with a place of nurture. She often remembered her grandmother sitting at the head of a bounteously spread table, dispensing her opinions along with the delicious home-grown food.

If grandparents provided recognition and security, aunts and uncle could contribute a touch of glamor. Robert Emmet Meriwether was Douglas and Carrie Meriwether's oldest child and only son. He was known as "Wild Rob," and his youthful adventures in Texas are the source of Caroline Gordon's short story, "Tom Rivers." His love of animals was proverbial, a passion Caroline Gordon shared. He could ride the wildest horses and gentle the most frightened beasts.[5] He loved children too, and always found time to give them their heart's desire, whether a ride or a story.

The aunts supplied the mystery of romance, past and future. Lucy Ann or "Loulie," the eldest daughter, had done the approved thing and married within the connection, and so retained the family name. This insular bliss was ended, however, when her husband, a doctor, became addicted to morphine. According to family legend, Loulie killed him by insisting that he break the habit by going cold turkey. Fortunately, many of her other attempts to help others were more successful, and she became known as the member of the family who would always shelter indigent, insane, or invalid relations, much like Caroline Gordon herself in later years.

Caroline remembered Loulie as having "the gait and mien of a lady abbess." The ecclesiastical rank was undoubtedly earned by Loulie's once having what is now known as an "out-of-the-body experience" when she began to die, but then returned to this world. She was greatly impressed with heaven, which she told the children was full of the latest labor-saving appliances. An electric egg-beater especially captured her fancy. Loulie left this paradise of domesticity because she said she had "to alleviate the lots of the rest of us." Like her mother Miss Carrie, she had the enviable gift of reading reality as she chose. With the onset of menopause, she claimed she had stopped menstruating intentionally because she saw that it was all foolishness.[6]

Loulie's youngest sister, Margaret or Piedie, was only in her early twenties when Caroline was a small child. She provided romance and drama because although she was not a beauty, she was a belle, and Merimont became a Mecca for her suitors. Gordon remembers that all the children enjoyed the sight of Piedie entertaining her beaux, and leading them on, in a swing in the back yard. She was "always in motion" and carried with her an "aura of excitement."

This "golden" world of Gordon's memoirs has considerable charm, especially for the modern reader, but it is a consciously created charm. From her memories and notes, Gordon selected those most in keeping with her chosen tone. At one point, this tone lapses when Gordon writes that "as a family, we were not given to celebration. Life was a grim affair, to be gotten through as best one could." Grimness, however, is exactly what she left out.

The family farms were not really insulated from the outside world. A neighbor and relation, Danforth Ross, comments, "After the Civil War, the days of affluence, or comparative affluence, were over for the Meriwethers and the Barkers. Few of them had been really rich, but they were comfortable. Now they were genteelly poor. They had plenty of food on the table but little money. The Civil War brought a mortgage to Merimont, and it stayed under mortgage in good times and bad." The aristocratic legacy had its disadvantages, according to Ross: "During the time I was growing up, the Meriwethers were regarded as too easygoing, not good business men." Their exclusivity could also have other connotations. Ross recollects, "My Uncle Ted blamed the Meriwether's problems on in-breeding. 'You take the Meriwether's and the Barkers,' he said. 'All crazy now. All run-down. That's what you get for being snooty.' "[7]

Similarly, the gentle nature of Caroline's grandfather, Douglas Meriwether, could be regarded as a fatal lassitude or passivity before the onslaughts of domineering Miss Carrie, who was capable of humiliating her husband before the hands for the way he cut the wheat.[8] Such an imposing and energetic spouse could only have aggravated the migraine headaches from which Douglas Meriwether suffered throughout his life. His death was caused by an overdose of the morphine he used to alleviate the pain. The family hoped the overdose was accidental, but some suspected otherwise. His last words express the frustration of futility, "If I could only know!"[9]

Since Douglas Meriwether died when Caroline Gordon was a child, Miss Carrie dominated Merimont, and continued to do so until her death in her nineties. Divide and conquer was her theory of power. She seated her family at the dinner table according to rank, their rank in her favor.[10] Discipline was also dispensed by Miss Carrie, as Caroline Gordon remembered.

> In the never-ending conflict between child and adult she represented authority. We enjoyed dwelling on instances of her tyranny. She kept a buggy whip lying at the foot of her bed—for our benefit she had told us more than once. True, she preferred giving an errant child a dose of 'peach tree tea' but if she was too tired to go out into the yard and break a switch off the nearest peach tree or was otherwise occupied, she would lean forward and lifting the buggy whip from its resting place at the foot of the bed, shake it to and fro, to let us know what she would do, if as she sometimes put it, we "pushed her too far."

"Peach-tree tea" was not Miss Carrie's only dose for children; she is remembered for her delicious sugar cookies that she always kept in good supply.

Miss Carrie's "tyranny" could be especially benevolent to a favorite grandchild, and she particularly favored those born on the place, such as Caroline Gordon, who was born on October 6, 1895, in the original, double-pen log cabin down the drive from the Merimont house. The central consciousness of Gordon's short story "One Against Thebes" contemplates the "Old Place" as Gordon might have. "She has been told that she herself was born in this house but she has no memory of ever having lived here. All her memories cluster around the tall, ugly gray house at the other end of the farm," her grandmother's house.[11]

Caroline also had the advantage of being her grandmother's namesake, and so she was Miss Carrie's acknowledged favorite. One cousin, Catherine Patterson Maccoy, remembers that Miss Carrie could look with pride on antics by Caroline that might have meant the switch for another child: a cousin "arrived and greeted my grandmother with the announcement, 'Carrie is up in one of your elms. She jumped up and caught a bough in the tree as we drove around the drive, and climbed up the tree.'"[12] Her cousin, playmate, and later sister-in-law, Polly Ferguson Gordon remembers that "whenever we, as children, wanted permission to go somewhere or do something, we would get Caroline to do the 'asking'—we knew the request would be more apt to be granted if she asked."[13]

Permission granted, the children enjoyed a halcyon world of their own within the larger adult world of the family farms. Caroline's playmates were her brothers and her cousins. Her brother Morris was three years older. He had the double distinction of being born on the place and being the oldest grandchild. Caroline remembers how he used this advantage by

> bestowing upon himself an appellation which, among country folks, is reserved for the good, the ancient, or the wise. He would refer to himself as "Your Uncle Dudley" . . . in the tradition of *Sut Lovingood* which *he* was allowed to read but *I* was forbidden to read because it was "too coarse." We children did not waste breath trying to convince him that while he was older than the rest of us he was neither conspicuously good nor wise. If he bullied us too much we complained to our father or mother or some other grown person. . . . Once when we howled louder than we had ever howled before, my father sized up the situation and ordered him not to refer to himself as Uncle Anybody. He thereupon begun speaking of "Your Non-Relative" and even went so far as to

inscribe cryptic and inflammatory messages from our Non-Relative on the door of the privy.

As his appellations suggest, Morris was a "bookish little boy" who once astounded a carpenter by reciting "Horatius at the Bridge" to him in its entirety.

Caroline's younger brother, William Fitzhugh Gordon, was quite different. He loved to play by himself in nature and showed no interest in letters. Caroline Gordon claimed that "Bill refused to read until he was nineteen years old and wanted to read detective stories." Their father "seemed actually pleased because my younger brother showed a positive disinclination to learn to read. 'William,' he said with a touch of pride,' is an atavism' " Caroline's affection for lively Bill is shown by making him her companion on the cross-lots, forbidden tour of the family farms that she recreates in her memoirs.

The masculine complement was further swelled by two orphan grandchildren of Great Aunt Cal Goodlett, Meriwether and Fletcher Baxter. Caroline remembers that Fletcher was her brother Bill's cohort in their many pranks. Meriwether had another distinction in her memory. "He was a well-built boy, with regular features and eyes that old ladies called 'a true violet.' He knew how to handle grown people and had a courtly way with people his own age. He was the only one of us who had attended dancing school and he was the first boy who ever kissed me. When I told him that I thought he was being silly, he looked surprised and calmly remarked that all the other little girls he knew liked to have him kiss them, as I am sure they did."

Caroline's closest friends were her Aunt Loulie's two daughters. Mildred Gordon Meriwether was "all rose and golden brown, and Marion Douglas Meriwether or "Mannie" was "blonde with eyes which were now gray, now green as her mood changed." The girls were slightly older than Caroline, and somewhat less interested in play, particularly tomboyish games in the woods. Mannie was "like a sister" to Caroline who kept in touch with her throughout her life. Their relationship is the prototype of the childhood friendship of Daphne and Agnes in Caroline's novel *The Women on the Porch* (1944).

Her cousin Polly Ferguson Gordon remembers games of paper dolls, hide and seek, spin the plate, drop the handkerchief, and musical chairs. Nearby Spring Creek was a favorite resort on hot afternoons. Polly Gordon also recalls that all the children "dearly loved to be read to," and when they weren't enjoying that pleasure, they were enacting what they had heard.[14] Two favorite stories were *The Last of the Mohicans* and *Robin Hood.* Caroline, as a girl, was often relegated by her brothers to what she regarded as the unexciting roles of the "elderly" Chingachgook of Cooper's novel and Allan-a-dale of the Merry Men. Her brothers consoled her for the latter role with the comment that Allan-a-dale was the only member of the band who could write.

Her brothers were canny in bribing her according to her bent, for Caroline was recognized as the "studious" one, the child who loved to read and write.[15] One of Merimont's remembered charms for Caroline Gordon was its significance as the place where her imagination was born and nurtured. The natural world beckoned the naturally observant child, but young Caroline was more than a passive spectator. "From early childhood any natural object, animate or inanimate, could arrest my attention, but I observed them, I now begin to suspect, because I wanted to tell somebody else how that tree stretched its branches or how that river flowed or that bird or beast moved." An account of her feelings on her fourth birthday demonstrates that her sense of vocation embraced its duties as well as its pleasures. "My 'work'—the work which, from earliest recollections I have felt called upon to do—came easily when I was quite young. The stories, then, seemed to take shape of their own accord. An imaginary person, doubtless a year or two older than I was and fairer in form and feature, wandered happily along the border of a sunlit wood. The adventures which I envisioned followed usually the same pattern, a pattern which I had no difficulty in apprehending."

This pattern came from three significant fairy tales that helped design her adult life and fiction as well as her childish fantasies. The first emerges in her account of learning to read.

> I do not know the day of the month or year when I learned to read. But I remember the moment. Some grown person—"Auntie," probably—had given me a bath before the fire in my grandmother's "upstairs"

> bedroom. After the bath she wrapped me in a big towel and had begun reading the story of Beauty and the Beast to me when somebody called to her from downstairs. She left the room, assuring me that she would be gone "just a minute." They tell me that I learned to read when I was four years old so I could not have been much more than four years old then. But I had already learned about time. I knew that a grown person's "just a minute" could last what was for me an eternity. I looked at "Beauty and the Beast" on the edge of a chair and I realized that an eternity might pass before my aunt might return. Propelled by a passionate desire to know what was going to happen, I tiptoed over to the chair where the book lay, standing on tip-toe, in my passionate desire to know what was going to happen, started reading.

Her fortuitous solitude and her readiness to read play their parts here, but the particular attractions of "Beauty and the Beast" are also significant. Although at the end of the story Beauty turns the Beast into a handsome prince with her kiss, much of the tale's suspense concerns the fate of Beauty's father. If one of his daughters does not take his place at the Beast's castle, the Beast will devour him. Beauty, of course, volunteers because of her great devotion to her father. Little Caroline, who worshipped her father, may well have identified with Beauty and longed for a similar opportunity to prove her love.

The other two fairy tales Gordon remembers from her childhood concern romantic rather than filial love. "The Sleeping Princess" or "Sleeping Beauty" and her entire family and court have been put to sleep for a hundred years by a spiteful fairy, not to be awakened until the Prince braves the tangles of the briar forest to claim the princess. As a young woman, Caroline Gordon was held by her love for her family in all its snarled, complicated branches amid the sleepy environs of Merimont, but she was also ambitious for life in a straightforward, more wide-awake world. For her, the Prince would be Allen Tate, who discovered her at her parents' house in Kentucky and led her to New York. In her notes for her memoir, her identification with Sleeping Beauty is clear. "The first 'ensnarlment' was resolved by the appearance of Allen on the scene."

Another favorite fairy tale taught her not to trust the first handsome prince who comes along because marriage may not be as pleasant as courtship. The heroine of "The Robber Bridegroom" is betrothed by her father to a good-looking, well-spoken young man who had happened by the house one day. To her horror, she discovers that he is really a robber who marries young women and cuts them to pieces in his house deep in the woods. She uses her wits to save herself when the young man arrives for dinner at her father's house and asks, "And now, sweetheart, do you know a story?" She tells what she knows about him in the form of a dream, leading him to betray himself in his astonishment and fear. Like the heroine of the tale who reveals truth in the form of a dream, Caroline Gordon often gave her version of the events in her highly autobiographical fiction. In the last novel she wrote before her divorce from Tate, *The Malefactors* (1956), she reveals what she believed to be some unpleasant truths about his character, and he, too, fled.

Curiously, these reminiscences about learning to read and write all cluster around her fourth birthday. In the story about learning to read, Caroline evades loneliness by immersing herself in "Beauty and the Beast." Her fourth birthday is also the date she gives for her putative "suicide attempt," when she found herself alone in her grandmother's room and plunged her face in the basin, rather than remain with the menacing presences in the shadows. The presences are not arbitrarily threatening, but associated with guilt since she had been "reflecting disconsolately that I had not done any work that day," meaning inventing stories. Art is work, but it is also an escape; it tells the truth, but in a transcendent form. The artist who does not create deprives not only the world, but herself, of truth and beauty. She who cannot face the truth kills her imagination in evading it.

In her unfinished memoir, however, Caroline evades much of the truth about her childhood by writing only of Merimont and its denizens. Her father, not the Meriwethers, truly dominated her childhood. James Maury Morris Gordon had all the charm of his literary avatar, Aleck Maury. In some ways a latter-day Huck Finn, he bewitches the reader with his ingenious ways of avoiding society's demands to pursue his true love, sport. "I want every day to be a pleasure to me" is Aleck Maury's most characteristic remark, and Caroline Gordon had also borrowed it from her father. While one can sympathize with James Gordon's desire to wring every drop of pleasure from fleeting time, the consequences were sometimes less than pleasurable for his family. Like Huck Finn, he did not care to stay in one place for very long, and he also indulged his impulse to "light out." After his children were born, he left his young family at Merimont and went to Europe for more than a year.[16] When he returned, the family moved on an average of every two years and Merimont became a summer refuge, not a permanent home.

There was probably more to her father's decision to leave Merimont than his undeniable urge to rove. In her notes for her memoir, Caroline Gordon wonders where she should discuss "enmity between my father and grandfather," Douglas Meriwether. In that highly inbred family, James Gordon had the disadvantage of not being a Meriwether, but he made more difficulty for himself by failing to conceal his scorn for the Meriwether's family pride. The remark of Sally Maury's father in "The Petrified Woman" could well have been spoken by James Gordon at a Meriwether family reunion: "All those mediocre people getting together to congratulate themselves on their mediocrity."[17] Caroline Gordon remembers that "all the Meriwethers" considered her schoolmaster father "a coward about heights and horses."

In the case of her grandfather, one can only speculate that his perceptions of his son-in-law's alien qualities may have been exacerbated by envy. Douglas Meriwether's childhood had been one of hard work and deprivation, as he helped his widowed mother support the family. From home, he proceeded to the Civil War, and after those hardships, attempted to rebuild the family fortunes by farming Merimont during Reconstruction. In *Aleck Maury, Sportsman,* Aleck repeats what he considers his father-in-law's most remarkable revelation. "He had once told me that he had never gone fishing except as a very small boy and had never had a gun in his hand until at the age of fifteen he enlisted in the Confederate army."[18] Perhaps being cast in the role of grasshopper to his father-in-law's ant was a factor in James Gordon's decision to leave the family connection.

Nancy Meriwether Gordon's role in these removals is more problematic, and indeed Caroline Gordon's mother remains something of an enigma. While at Merimont, Nancy Gordon would always be ruled by her mother, forceful Miss Carrie, as a favorite family anecdote illustrates. When Nancy Gordon had her first child, Morris, at age eighteen, the black nurse called out to Miss Carrie that she'd better come quick because Miss Nan was holding the baby and would surely drop it. Even after the Gordons' move away from Merimont, to Hopkinsville, Kentucky, Miss Carrie still wanted to raise her grandchildren herself. In a letter to a friend on February 9, 1902, Nancy Gordon wryly comments, "As for Carrie, well, if I ever have another child I'll not name it after anybody. She has been at home ever since Christmas, and on account of the weather I have been quite helpless. Mamma's reasons are very touching as to why it is utterly impossible to send Carrie."[19] Several years later, when her husband wanted to return to Merimont after living in Clarksville, Nancy Gordon refused.

Aside from her need to escape her domineering mother, Nancy Gordon probably wanted to leave Merimont because of other differences with her family. She was passionately intellectual, a "bluestocking," and a "linguist." She was known for cooking supper with a ladle in one hand and a French novel in the other.[20] When at the age of seventeen she married her teacher, she may have been seeking what the farm-and-family obsessed Meriwethers were unable to offer, a world of the intellect; her daughter Caroline would leave the family connection on a similar quest. Polly Gordon remembers Nancy Gordon as "a deeply religious woman" who once "worked (teaching) and saved one thousand dollars which she gave toward the support of a missionary."[21] Nancy Gordon herself wrote of the Meriwethers, "All of my family are so prejudiced against the church and regard it as a very worldly institution, and rather to be avoided from a religious standpoint."[22]

Nancy Meriwether may have hoped that by marrying James Gordon she could avoid conforming to Meriwether ways and profit from the companionship of a man who shared her intellectual interests and was sympathetic to her religious beliefs. As in a fairy tale, she was granted her wishes, but with some twists that made the situation repugnant to her. Instead of nursing her intellect, she nursed her husband's valuable but motherless setter puppies.[23] She found that his obsession with finding new territories for sport not only kept them away from Merimont, but had them moving for the rest of her life. Caroline Gordon remembered her mother as "withdrawn" because "she didn't like the way we lived."[24] Nancy Gor-

don had fled from a domineering mother into the arms of a domi-
nant male, and she still could not control the way she lived.

The tensions between James and Nancy Gordon had important
ramifications for their relationships with their children. Morris, the
oldest and the first grandchild, was a sort of universal favorite with
the eldest's child's preeminence. Bill was "the apple of his mother's
eye." Carrie was her father's favorite child. She looked like the
Gordons and sided with her father against her mother. In her memories of growing up, she stresses how much she learned from her
father's schooling in composition and languages, but oddly does not
mention any tutelage from her intellectual mother. Similarly, Caroline acknowledges her father's example as a magnetic teacher, but
he was not the sole influence on her distinguished teaching career.
In Clarksville, Nancy Gordon is still remembered as a talented
teacher at the South Kentucky Bible College and later at her husband's school in Clarksville. Interestingly, like her mother, Caroline Gordon also became devoutly religious, in her later years.

Caroline Gordon was as much her mother's daughter as her
father's favorite, but she believed her mother did not love her,
perhaps because of her resemblance to her father or her mother's
favoritism toward the youngest child, Bill. When her mother was
discussed, the adult Caroline Gordon would always tell this anecdote: when she asked her mother if she loved her, she replied,
"Carrie, I will always do my duty by you." Her father apparently
did not help matters any, as in the other parental precept she
related. Her father told her that if her mother said black was white
or white was black, she was to agree, as he did, to keep the
peace.[25] His policy may have kept the peace, but it was the stillness
of alienation and failure to communicate.

Whatever their differences, James and Nancy Gordon did have in
common their vocation as teachers, and they increasingly shared a
religious call. While at the South Kentucky Bible College in Hopkinsville, James Gordon began to preach on Sundays at churches in
Madisonville and Pembroke. His denomination was "Campbellite,"
more formally known as the Disciples of Christ or the Christian
Church. The sect was founded by Alexander Campbell (1788–1866)
in 1832 when he broke with the Baptists. Although he shared their
belief in baptism by immersion, he had some disagreements with

them. The Campbellites opposed both "speculative theology and emotional revivalism." In Bethany, West Virginia, in 1840, Campbell founded Bethany College, from which Caroline Gordon would graduate in 1916.[26]

James Gordon's motives for becoming a Campbellite preacher remain a subject of speculation. One reason proffered is that the life of a minister allowed more time for sport than that of a schoolmaster.[27] Caroline Gordon left another explanation among her notes:

> During the summer vacations he fell under the spell of my grandmother's father, John D. Ferguson, a Campbellite preacher. . . . I once asked my father what Grandpa Ferguson looked like. He thought a minute and said, "He must have had dark hair when he was young. His hair was white when I knew him." He added, "He was the best man I ever knew with the prepositions." The prepositions were Greek prepositions on [the correct translation of] which points of Campbellite doctrine depended. My father turned Campbellite preacher and devoted his not inconsiderable powers to homiletic exercises and a study of the Typology of the Scriptures. His sermons were marvels of conciseness and always homiletically sound, with the poetic insights that mark the true theologian. They should have been preserved.

Her account, written in the last years of her life, makes her father into an anachronistically "poetic theologian," according to the precepts of Jacques Maritain. She also praises qualities that characterized her own prose: conciseness, poetic insight, and an almost obsessive need for correctness in the smallest detail. Caroline Gordon's story also leaves out the influence of her religious mother; indeed it eliminates her from the genealogy since Ferguson is called "my grandmother's father."

James Gordon continued his combined career of teaching and preaching when he decided to found his own college preparatory school, Gordon's University School, in Clarksville, Tennessee, not too far from the Meriwether family farms. In the early years of this century, Clarksville was a world center for the marketing of a heavy, dark tobacco popular in Europe. The Gordons hoped to find pupils in this prospering city, and so they settled at a two-story brick house on Madison Street across from the entry of Seventh Avenue. The family lived upstairs, where Nancy Gordon also taught some of the younger boys and a few girls. James Gordon taught in a long front room downstairs, a combination of two adjoining rooms.[28]

To this day James Gordon is remembered as a strict disciplinarian who would punish unruly boys with the strap or simply expel the incorrigible. In his ungraded classroom, in some ways reminiscent of today's open classrooms, decorum reigned so that all the boys could master their lessons at their own pace. James Gordon, however, did not suffer fools gladly. He could deflate the pride of a doting mother by inquiring, "Mrs. So and So, you are a very ordinary woman and I have never thought Mr. So and So had more than average intelligence. What makes you think Guy will amount to much?"[29]

His strictness was respected because of his devotion to learning. He was remembered as an inspiring teacher, particularly of English literature, which was then considered an innovative addition to the standard curriculum of the classics and mathematics. He would not allow the boys to read aloud, however, since their mistakes and hesitations could spoil the beauty and sense of the passages. Instead, he read to them and kept the boys spellbound with his mellifluous deep voice. Stern though he could be, James Gordon was kind and generous to any boy who was truly interested in learning. One of his pupils, Oscar Beach, remembers that Professor Gordon did not charge his mother tuition for him because he knew the young widow with four children could not afford it, and he also knew how much young Oscar wanted to learn.

These years in Clarksville were apparently happy ones for Caroline Gordon. In *Aleck Maury, Sportsman,* she remembers them with vivid and accurate details, sometimes changing only the names of people and places, and occasionally not even providing that disguise. In her short story, "The Petrified Woman," her fictive persona, Sally Maury, reminisces about her father. "He is Professor Aleck Maury and he had a boys' school in Gloversville then. There was a girls' school there too, Miss Robinson's, but he said that I wouldn't learn anything if I went there till I was blue in the face, so I went to school with the boys. Sometimes I think that is what makes me so peculiar."[30]

One such peculiarity could be Gordon's lifelong distrust of the quality of women mentors, and women's abilities in general, and her reliance on a series of male mentors, beginning with her father.

She did, however, gain a solid foundation of knowledge from James Gordon. By age eight she was reading Caesar's *Gallic Wars.* Her study of Greek grammar began when she was ten. Her father was also her first teacher of what we would now call creative writing. His praise, "Daughter, it's so darned good!" and his criticism, "That is precious," were her touchstones. Unfortunately, none of these early efforts seem to have survived.

Caroline Gordon remembered that "we were more prosperous, when we lived in Clarksville, then we had ever been before or would ever be again, two horses and two servants." This period of prosperity came to an end after several years, but not, however, from any failure of the popular and respected school. James Gordon was still trying to combine three "vocations": preaching, teaching, and sport. He relegated his preaching to Sunday sermons in local churches, and he dismissed his school at two o'clock every afternoon to take to the fields and streams. Despite this schedule for making every day a pleasure to him, James Gordon was beginning to feel bored and confined after several years in the same place. Caroline Gordon wrote, "My father, meanwhile, murmured darkly at intervals about the state of his health. He often had dizzy spells, in the school room of course, never in the field or beside the stream. He sometimes spoke of what a catastrophe it would be if he should measure his six feet on the school room floor."

To avoid such an untimely end, James Gordon decided to go into the ministry full-time. He could devote an hour or so a day to his sermons and then pursue sport. He is remembered as "quite an orator," so those morning hours were apparently sufficient to their purpose.[31] His family, however, found themselves once more on the move, and these moves are difficult to date with any accuracy. Sometime after Clarksville, he took his family to Wilmington, Ohio, where Caroline Gordon said she attended school formally for the first time, the local high school. She remembered the time at Wilmington as about four years, but we do know that by 1910, when Caroline was fourteen, James Gordon was ready for another move. On February 28, 1910, the local newspaper, the *Clinton Republican,* contained the following item about James Gordon's sermon the preceding day:

> In a touching address he expressed his sorrow at leaving his very happy associations here, but stated that he desired to go with his family to farm life. Rev. Gordon has lately purchased a farm 5 miles from Lynchburg, Virginia, and will move his family there as soon as a successor can be secured for his pulpit. Mr. G. will become the State Evangelist of his church in Virginia, and will give his boys a chance to work at farming. His congregation learned of his intention with regret, for he had greatly endeared himself to all his people. Dr. G. is a scholarly gentleman, and had made a lot of friends here, who will be sorry to lose him.

Presumably, an evangelist would have even fewer settled duties than a pastor, and could allot his time more freely between the duty of proselytizing and the pleasure of sport.

Little is known about this time in Virginia. Those who knew Caroline Gordon well, including her daughter, do not recall her mentioning it. The only reference I have found in her papers is "Virginia" listed *before* "Wilmington" in the outlines for her memoir, but Caroline Gordon was generally quite vague about dates and sequences. Her cousin Polly Ferguson Gordon remembers that James Gordon did not succeed at farming in Virginia, and so returned to preaching in Kentucky.[32] Perhaps this Virginia venture is what Caroline Gordon had in mind when she wrote that her father always liked to farm, on paper. At any rate, by 1911 he had proceeded to pastorates in or near Princeton and Madisonville, Kentucky, as Caroline Gordon's college transcripts note. By 1915, the peripatetic James Gordon had moved his family once again, this time lured by the fishing of Poplar Bluff, Missouri.

In the midst of these peregrinations, Caroline Gordon was completing her formal education. To be admitted to Bethany for the fall of 1912, she submitted a "Certificate of Preparatory Work" from the Princeton Collegiate Institute, presumably meaning Princeton, Kentucky, since her father's ministry was near there.[33] The certificate spells her name as "Carolyn," which is the way she spelled it throughout her youth. Since the spaces for times of attendance are left blank, she may not have been a pupil at this institute. The signer, the principal "Mrs. Miller," may have examined her and then attested that Caroline had accomplished so many years of academic work in various disciplines, since such certification was a possible route to admission at Bethany. Caroline Gordon's daughter remembered that her mother had mentioned attending a finishing school, where she learned to draw, before she matriculated at Bethany.[34] The certificate does attest to her abilities in crayon and watercolor, so it is possible that she attended this institute, however briefly. According to the certificate, she was also proficient in English literature, Latin, German, history, algebra, plane and solid geometry, and physics. Her best subjects were listed as Latin, Greek, German, and Literature.

The choice of Bethany College was not at all surprising. Bethany, West Virginia was not too remote from her Kentucky home, and was easily accessible from Wheeling by a trolley, the "Mountain Canary." In fact, part of her home had preceded her there in the form of the many relations and neighbors who had attended Bethany. While Caroline was at Bethany, her favorite cousins, Mildred and Marion Meriwether, were also students.

Most important, Bethany was a Campbellite college, where the "atmosphere" was "unusually healthful and stimulating," according to the College Bulletin of 1913. Daily attendance at chapel was required. Moreover, the environs of the town of Bethany were "free from saloons, wicked resorts and other influences all too common in other college towns." The tone of the college and its sense of mission are well expressed in a passage from the 1913 Bulletin that finds metaphorical significance in Bethany's physical setting overlooking the Buffalo River. "The college is the mountain summit whence streams flow down upon the fields of life. What is being taught and thought up there will presently appear as a practical force down on the level of character and conduct. If agnosticism takes possession of the mountain summits, religious faith and faithfulness cannot hold the plain." Aside from the lure of its embattled Christianity, Bethany may also have been attractive to the Gordons because it offered a one-third reduction of the tuition to the children of ministers.

The college had more secular attractions as well. The physical plant had been much expanded by recent building, including a new library for which Andrew Carnegie had provided the funds. Phillips Hall for Young Ladies, where Caroline Gordon presumably resided, is described in the College Bulletin as a "stately and commodious building, with all modern improvements, heated with steam, rooms newly carpeted and furnished with comfortable and substantial hardwood furniture, well lighted, well ventilated, and altogether adapted to the needs and conveniences of young ladies." The college seems to have been rigorous in its academic requirements, stipulating class attendance, public orations, and quite a range of courses.

Caroline was somewhat apprehensive about her preparation for college, since she had so little formal schooling. Her anxieties were reinforced by her mother.

> My mother, having married so early, had never had an opportunity to go to college. She was a blue-stocking, nevertheless. I think she may have envied me when I went off to college at the age of seventeen [the age at which Nancy Gordon had married]. However, she had done her best to prepare me for the momentous event. Early in the summer she had handed me *White's Greek Grammar,* telling me that it was my task to master it before the term opened. I don't think I missed a dance or a swim that summer but I did "bone" away at odd intervals until the end of the summer when, just as I reached the protases and apodoses, my mother informed me that in addition to mastering White's First Greek, I must translate the whole of Zenophon's *Anabasis.* I was not particularly disconcerted by this news. I had been reading excerpts from the *Anabasis* all summer in my grammar and had got so I rather liked the stuff.
>
> Yet I was in for a surprise when I reached the little college tucked away in the hills. My mother had mis-read the catalogue. The requirements for first-year Greek were much less stringent then she, in her innocence, had supposed. When my faculty adviser found that I had had some grammar he put me in third year Greek, with three or four "sharks." We read the Odyssey the first semester, Thucydides *Peloponnesian War* the second. The sharks dropped out after that. They had had all the Greek that was "required."[35]

This anecdote attests to Caroline Gordon's devotion to her studies and her determination to master every detail, "required" or not. Here Caroline mentions her tutelage by her mother, instead of her father, though her mother is portrayed as a critic who makes unnecessary demands through a "mis-reading" of the text.

In her teacher of Greek at Bethany, Professor Frank Roy Gay, Caroline Gordon found a mentor to replace her father. Gay received his bachelor's and master's degrees from Drake University, and did additional graduate work at the University of Virginia and the University of Chicago. He taught at Virginia Christian College in Lynchburg from 1907 to 1910, so it is possible that he met the Gordon family when they first arrived in Lynchburg in 1910.[36] In 1913 Gay joined the faculty at Bethany. He was noted for his devotion to the study of the classics, of which he wrote that "nothing has a greater power to develop the judgment and to purify the tastes."[37] In fact, his zealous study made him somewhat unworldly, as is attested by a popular Bethany anecdote about him. Professor Gay kept his research on note cards at home. When his house was burning down, his first cry was supposedly, "My cards, My cards!" and then, "My wife, My wife!"

Although women may not have been first in Professor Gay's priorities, he respected Caroline Gordon's abilities, and she performed prodigious feats of scholarship to retain that approbation. After she completed the third-year Greek class in her freshman year, Professor Gay decided to offer Greek IV, the drama, although Caroline would be the only pupil. Professor Gay's time was certainly not wasted, for Caroline Gordon later stated that Greek IV gave her "more practical knowledge of the craft of fiction . . . than anybody was ever likely to acquire in one of my classes in creative writing."[38]

In Greek drama Caroline was drawn to qualities that would later mark her own fiction, such as the importance of genealogy or "blood" in accounting for a character's personality and actions. In his explication of the *Agamemnon,* Professor Gay provided another touchstone for the future novelist, the description of the carpet on which Agamemnon walked to his death. "He affirmed that if that carpet had not been laid down—so firmly that one has the illusion that one can see the very threads of which it is woven— that Aeschylus could never have achieved the superb irony which plays like summer lightning over the whole action."[39] Caroline would eventually achieve a similar ability to endow a meticulous, realistic description with richly symbolic overtones.

Another harbinger of Caroline's later practices is exhibited in this anecdote about Greek IV.

> I flunked trigonometry and calculus and would have flunked physics if a boy friend hadn't written up my experiments for me. But I never neglected to translate my 100 lines a day—except on one occasion when I had stayed out late at a party and early the next morning rushed around

to the library and provided myself with a "trot." "Miss Gordon," my mentor said kindly as I was attempting to translate a passage from the *Antigone*. "I hesitate to say that you are wrong, but the weight of opinion is against your rendering of *kata* in that context. Jebb is the only authority who ascribes that particular meaning to it. I took Sir Richard Jebb back to the library and never consulted him again. I was convinced that where the tragedians were concerned my professor had second sight.[40]

In using her "boy friend" and Sir Richard Jebb as uncited sources, Caroline revealed her need to ward off criticism by always producing the correct details or answers. She always felt fundamentally unprepared and in danger of being "found out," perhaps because of her lack of formal education to this point. She is also, however, demonstrating the insecurity about her own abilities and the reliance on masculine authority that led her to quote passages from documents and historians in her historical fiction without any indications that they are quotations.

This anecdote also gives a glimpse of the lighter side of Caroline's years at Bethany. Despite her serious attitude toward her studies, she did attend parties and visit the homes of her classmates. According to the college yearbook, the *Bethanian*, she was a member of the Alpha Xi Delta sorority; the Merry Maskers, a drama group; a literary society, the American Literary Institute; the YWCA; and the *Bethanian* staff. The Senior Class Prophecy hints at one boy friend, but not her future distinction, when it foresees "the President of the American Tobacco Company, A. E. Sims, with his wife, Carolyn Gordan [*sic*] Sims." Like her father, Sims was from Louisa County, Virginia; his membership in the Ministerial Association indicates a similar vocation; and his literary interests are attested by his membership in the American Literary Society.

Alfred Sims, however, was not the young man Caroline Gordon recalled in an interview when she was asked about her college romances. She said that she had a beau named George, who was expelled from Bethany for criticizing the college in the newspaper. George wanted to marry her and take her out West, but she refused him because she was afraid. Laughing, she would not reveal his complete name, but her notes mention a "G.A.H." in her

Bethany years.[41] Under her picture, the 1915 *Bethanian* states,
"Although she is a true Southerner, she is evidently cosmopolitan
in her views, for she has developed of late a remarkable fancy for
Colorado." The 1915 *College Bulletin* lists a George Archibald
Hankins of Pueblo, Colorado, who does not reappear in the 1916
edition. Whatever the actual identity and history of "George," her
memory of him indicates that she was attracted to rebellious young
men who wrote against authority. Nearly a decade later, when she
met Allen Tate, a contributor to *The Fugitive* and champion of
modernism, she would find the courage to let him take her away.

Although she had rejected George's plan for her life, she did not
really have one of her own when she graduated from Bethany in
1916. She was, however, qualified to teach since she had also
acquired a degree in Pedagogy at Bethany by the end of her sopho-
more year. In an interview she said that from her graduation in
1916 until she began work in Chattanooga in 1920, she went home
and "fooled around" and taught at a co-ed high school in Clarksville.[42] That is all she ever seems to have said about this four-year
period, in print or to her friends and relatives. Her daughter,
Nancy Tate Wood, speculates that her silence indicates that she
was miserable because she wanted to get away from her family and
write.[43] Perhaps she did not wish to emphasize the length of her
apprenticeship since her first novel did not appear until 1931, or
perhaps she merely wished to minimize the fact that she was four
years older than Allen Tate.

What little information we have about these years comes from
the records of the Clarksville High School and the memories of two
of her students. The principal's report for 1918–19 indicates that
Caroline Gordon had three years of teaching experience, including
that year at Clarksville.[44] The other two years were at Putney
Normal, but its location is not mentioned. Since Caroline's parents
lived in Poplar Bluff, Missouri, when she graduated from college,
it is possible that her early teaching experience was gained in or
around the town.

A student in her beginning French class at Clarksville, Anne S.
Major, provides a portrait of the artist as fledgling teacher.

> She was an attractive, intelligent young woman, performing what must
> have been a chore, and I was a timid rather frightened pupil. She was

small in stature, slim and quite trim in figure. Her hair was worn up, not cut, somewhat fluffed around her face. Her eyes and brows were the most outstanding features of her face, dark brows almost meeting. I also liked her manner of speaking which was clear, interesting to my ear, and almost foreign. I never addressed her directly in my whole year in her class, but for some reason I was always attentive to each brief encounter in and out of the classroom. Her dress was always plain, appropriate, and usually black in color, straight in fashion.[45]

The quiet, understated, almost classical elegance would remain characteristic of Caroline Gordon, as would the qualities that made her mesmerizing and inspiring as a teacher. It is not surprising that Anne Major decided to become a teacher of French at Clarksville High School.

In these early years as a teacher, Caroline Gordon's sensitivity, insecurity, and self-consciousness sometimes obscured the master teacher she was to become. At Clarksville, she taught science, French, and agriculture. Although her stays at the Meriwether farms may have helped with agriculture, she always had difficulty in learning to speak foreign languages and displayed little interest in science. Perhaps this strange conglomeration of courses helped contribute to the discomfiture remembered by Frances Moore, a student in her senior science class. "One day while we were working in the chemistry laboratory, she sat at her desk and cried all during the class period. We were so surprised to see a teacher cry in class, that we quietly worked and did not ask a question."[46]

Anne S. Major remembers another instance of the young teacher's embarrassment and distress.

One afternoon . . . the girls in my French class gave a program in the assembly that represented a take-off on all women faculty members. Iola Gracey Smith was selected to represent Miss Gordon, penciled brows, straight black dress, rather typical and recognizable. The feature that hurt our young teacher was the posture Iola assumed in her manner of sitting, either knees apart or too much leg showing. Laughter from the audience and anger and tears from the real Miss Gordon. In those days, she was a truly discreet and sensitive person.[47]

The little girl who had cried because the photographer said she was less attractive than her brothers was still alive within the attractive young woman hurt by student silliness.

Another incident attests that however self-conscious she might be, Caroline Gordon never concealed her affection for those she loved, no matter how they appeared to others. Anne S. Major writes,

> One other experience left in my mind of this teacher, is the day I arrived somewhat ahead of the class. Miss Gordon had a vacant period just prior to our French class, and on this day she had been entertaining a visitor who was on the point of departure. I entered, took my seat, and saw at the front of the room a man standing with his back to the classroom, facing my teacher, almost her same height but not obliterating her small figure. They continued to speak quietly as the class assembled; the farewell was with an embrace and a kiss. In my snobbish school girl way, I was astonished. The man, somewhat older than Miss Gordon, wore a plain drab brown suit, a gray faded felt hat, giving the appearance of the country man, a farmer perhaps in his town clothes. As I thought about this brief scene, I became proud of her and convinced of warmth I had never suspected.[48]

Whether this man was Meriwether kin or a local beau remains unknown, as does much of the fabric of Caroline Gordon's life in this period.

By 1920 Caroline Gordon was ready for a change. Her cousin Polly Ferguson Gordon remembers that "Caroline spent most of the summer of 1920 at 'Merimont'; so did I. It was a time of decision for her—she was with her parents some of the time. She did not teach during that time. I expect she was pondering her career and made the decision to write."[49] Teaching is a notorious drain on creative work, so Caroline may have been seeking a way to devote her best energies to writing. She also may have been trying to extricate herself from her tense relationship with her mother. An anecdote Caroline told her daughter probably dates from this period. A man wanted to marry her and sent telegrams, but her mother never gave them to her, and so she lost him.[50] Whatever the facts of the case, Caroline's resentment of her mother is clear. The simplest explanation may be best: at age twenty-five, Caroline Gordon may have decided it was time to leave home and make her own career.

The extensive Meriwether connection intervened once again, and she found herself living in Chattanooga with her mother's sister, Margaret Meriwether Campbell, whom she called Aunt

Piedie, as Piedie called Caroline, "Kidie." In her aunt's attic, she wrote her first novel, *Darkling I Listen,* which she later destroyed unpublished. According to Caroline, it was highly autobiographical, as first novels often are. She said that she wrote the novel in a nervous depression and credited her Aunt Piedie with saving her life during this breakdown.[51] The stanza of Keats's "Ode to a Nightingale" from which the title was taken is indicative of her feelings at the time.

> Darkling I listen, and for many a time
>   I have been half in love with easeful Death,
> Called to him soft names in many a mused rhyme,
>   To take into the air my quiet breath;
> Now more than ever seems it rich to die,
>   To cease upon the midnight with no pain,
> While thou art pouring forth thy soul abroad
>   In such an ecstasy!

Caroline was experiencing the beginning writer's intense frustration at failing to express herself, yet for her words seem to be the only alternative to death. She found that writing was an art to be mastered; words did not pour forth with the natural ease of the nightingale's song. Since she had not yet encountered a sympathetic mentor, she had little confidence in her talent and even less in her knowledge of technique.

Despite her despair over her fiction, Caroline did make significant advances in her four years in Chattanooga. Although she lived in her aunt's house, she was independent enough to support herself since she found a job as a reporter for the Chattanooga *News.* In her study of Gordon's fiction, Rose Ann C. Fraistat evaluates her apprentice work: "One of Gordon's first responsibilities . . . was reviewing new books. For the early months of 1920, her column appeared weekly under the by-line 'Carolyn Gordon.' Written in the more casual tone and with the brevity expected in newspaper reviews, her criticism reveals, nonetheless, the appreciation of craft which marks her later essays."[52] This attention to technique would be characteristic of her fiction and teaching as well.

Her weekly reviews forced Caroline to become aware of a broad range of literature and to gain a sense of the world of letters, including the beginnings of modernism. One of her assignments as a features reporter was an article about the various little magazines that were springing up in the area, including *The Fugitive,* the product of some Vanderbilt faculty and students. As she praised *The Fugitive* in her article of February 10, 1923, she was inadvertently introducing herself to her future husband. In *The Fugitive Group,* Louise Cowan quotes a letter John Crowe Ransom wrote to Tate the next day: "When I get copies tomorrow I'll send you one. . . . Written by a Miss Gordon, who has developed quite a fondness for us, and incidentally is kin to some of my kinfolk in Chattanooga."[53]

Caroline Gordon and Allen Tate did not actually meet until the summer of 1924 in Guthrie, Kentucky, where Caroline's parents were living at the time and she herself was home for a vacation from Chattanooga. Tate was visiting his good friend and fellow Fugitive, "Red" Warren, later better known as the poet and novelist Robert Penn Warren. One day Caroline's father encountered Tate and Warren on the street and suggested they come home with him and meet his daughter who was as "crazy" as they were. Caroline's father was right about their affinity, if not necessarily its cause.[54] The attraction between Allen and Caroline was instant, and a passionate love affair ensued. Caroline Gordon now knew what she wanted to do with her life: join Allen Tate in the cosmopolitan world of literature. As one of her favorite fairy tales had predicted, the Sleeping Princess had found the prince who would lead her to her destiny.

CHAPTER 3

Ironically, the prince who was to rescue Caroline Gordon from her "ensnarlments" had a similar feeling of entanglement with the branches of his family tree, but lacked Caroline's sense of rootedness. Like Caroline, Allen Tate imaginatively identified with his mother's side of the family. In a poem beginning "Maryland, Virginia, Caroline," he linked their home states with the name of his wife as the "source" of his "farthest blood."[1]

Tate's mother, Eleanor Parke Custis Varnell (1865–1929), was connected to the Bogans of Fairfax County, Virginia, through her mother, Susan Armistead Bogan. Eleanor Tate endowed the young Allen with a strong identification with his Bogan ancestors by taking him on frequent pilgrimages to the sites of their old Virginia homes at Pleasant Hill and Chestnut Grove. She even told him that he was born at Chestnut Grove, and he did not learn otherwise until he was thirty. Tate himself recognized her wish to align herself with the Bogans when he wrote, "The house where she might have been born, Pleasant Hill, was burnt to the ground by General Blencker's New York 'Dutch' Brigade on July 17, 1861."[2] Thus, though neither Tate nor his mother was born in the "right" place, she managed to replace the Tates with the Bogans as Allen's imaginative fathers. Major Lewis Bogan (1795–1870) became the model for Major Lewis Buchan in Allen's only novel, *The Fathers,* although his link with him was through "the mothers."[3]

The material comfort and security of Eleanor Tate's life came from her father, George Henry Varnell (1833–1899) of Maryland who made his fortune in timber and land. Indeed, he even provided for her after her marriage, by insisting that her unemployed spouse, John Orley Tate (1861–1933), take a job with his lumber company in Illinois. Even before his union with family-conscious Eleanor Varnell, John Tate was at something of a disadvantage since he was an orphan, raised by his maternal grandfather. His lack of parents, however, did not mean he lacked a worthy lineage of his own. His father, James Johnston Tate (1819–1872), was from a planter's family in Chester County, South Carolina. He became a schoolmaster to the Allen family of Shelby County, Kentucky, and married one of his students, Josephine Allen (1840–1867), a curious prefiguration by Tate's paternal grandparents of the circumstances uniting Caroline Gordon's parents. James Tate's Carolina background and Kentucky home also reverberate in Allen's ancestral triumvirate of "Maryland, Virginia, Caroline," later reinforced by his wife Caroline's own Kentucky background.

John Orley Tate and Eleanor Varnell had three children: James Varnell was born in 1888, followed by Benjamin in 1890, and, nearly a decade later, by John Orley Allen Tate, known as "Allen," on November 19, 1899. By this time the family was back in Winchester, Kentucky, where John Tate supported them by a lumber business, land sales, and stock speculation. He was successful in none of these enterprises, and by 1920 the family's maintenance was in the hands of his son Ben, a successful entrepreneur.

Long before the financial failure was acknowledged, however, marital bankruptcy had been tacitly declared, and Allen's parents lived apart through much of his childhood. The reasons for the Tates' separation were not evident to Allen, and he later wondered about them in his memoirs. He states that his brother Ben told him his father was involved in some kind of sexual scandal and "had been asked in 1905 to resign from a men's club, having laughed at the mention of a certain woman's name."[4] Allen was not even sure which parent "left" the other. As an elderly man, he wrote, "I still do not know whether my mother's perpetual motion was flight from her husband, or whether his long absences were flights from her."[5] One might have predicted alienation in the union between a woman so proud of her active and prosperous relations and a man who had no inclination or aptitude for work and was subject to spells of lassitude, much like those Allen would later suffer in his own career.

Whatever the reason, the result was clear: Allen Tate became his mother's companion in her migrations through the South, from residential hotel to mountain resort, with some pauses at the family home in Winchester, Kentucky. Allen wrote, "We sometimes moved two or three times a year, moving *away* from something my mother didn't like; or perhaps withdrawing would be a better word; for my mother gradually withdrew from the world, and withdrew me with her."[6] Like Caroline Gordon, Allen perceived the results of parental discord as maternal "withdrawal." Unlike Caroline, Allen did not become aligned with his father and alienated from his mother; he was not only by his mother's side, but he took her side against his father. "When I was about eight years old in Nashville," he remembered, "one morning a letter arrived that upon reading my mother burst into tears and left the room. I seized the wicked paper and threw it into the fire."[7]

Such humiliation, puzzlement, and rejection were typical of Allen's attitude toward his father, as another memory illustrates. After his promotion to long pants, Allen began to steal some whiskey from the sideboard and replace it with water. In his memoirs, he describes his father's reaction to the pilfering.

> He was a large, handsome man, about six feet, whose eyes were blue, cold, and expressionless. I was afraid of him; I never stopped being afraid of him, for he always spoke to me impersonally as if he were surprised that I was there. He gave me a fishy stare, straight through me, and said, "Son, the next time you steal the whiskey, don't ruin it for other people by filling up the bottle with water." There was a moral in it, but I didn't know then and I am not sure now what it was: possibly that only low whites and Negroes stole whiskey and that drinking must not be secret. It, too, was a public ritualistic act, and it followed that I had to be humiliated before company at dinner.[8]

Allen's speculations evade the motive his account subliminally conveys. His father was telling him that he, and his morals, were of no consequence, not even worth anger. Indeed, Allen fears for his very identity since his father does not even acknowledge his existence, looking "straight through" him, and acting "surprised" at his presence. Allen later overcame his fear by adopting his father's manner, so that those who were not his intimates would comment on his coldness, arrogance, and seemingly Olympian detachment. If the story about the elder Tate's sexual proclivities is true, Allen also learned his later pattern of philandering from his father. In short, his resentment toward his father may be a projection of his discomfort with his own least admirable traits of character.

Indeed, even his father's home base, the environs of Winchester, Kentucky, are connected with ugly emotions in Allen's memoirs. He mentions "the lynching I had seen when I was eleven," and on the same page recounts his own act of racial scapegoating in Winchester: "there I had let a Negro boy, my playmate, take a beating from his mother, our cook Nanny for a petty theft I had committed. (Henry was killed in World War I; I never made it up to him for my cowardice.)"[9]

Although his father's chill detachment aligned Allen with his mother, he was ambivalent toward what he regarded as her overwhelming influence on the formation of his intellect. He was educated at home until age nine, but the boy who was to become a prominent intellectual was not regarded as bright. "One day when I was about twelve, leaning against the newel post in the front hall . . . , reading *Through the Looking-Glass,* or perhaps *Miss Minerva and William Green Hill,* my mother said, 'Son, put that book down and go out and play with Henry. You are straining your mind and you know your mind isn't very strong. So, at the age when most boys turned to backflips to impress the girls, I was quoting poetry to them as surrogates to my mother, to whom I had to prove I was not an imbecile."[10] Tate is suggesting the source of his fierce intellectual competitiveness, which emerged as the brilliance that dazzled many women, including Caroline Gordon.

Later, while writing for *The Fugitive* at Vanderbilt, he adopted the pseudonym "Henry Feathertop," an allusion to Nathaniel Hawthorne's short story by that title. Mother Rigby, "one of the most cunning and potent witches of New England," decides to make a scarecrow for her corn patch which "should represent a fine gentleman of the period."[11] She constructs him of broomsticks, old clothes, and furniture, stuffs him with straw, puts her dead husband's wig on his head, gives him a pipe of tobacco for the breath of life, and thus has created what appears to others a living, breathing, moving human being called Henry Feathertop. The scarecrow, however, realizes his falseness and destroys himself, crying "I've seen myself, mother! I've seen myself for the wretched, ragged, empty thing I am!" The persona of Allen's early poetry often refers to himself as constructed of discarded pieces of the past which are "mildewed" and "withered," culminating in his self-portrait as "a mummy, in time" in his "Ode to the Confederate Dead." Although Allen is ostensibly referring to the burden of the southern past, in terms of his own psyche he may well have feared that "mummy" had created a mummy.

The tale contains another passage that may have been significant for Allen. Hawthorne writes that the scarecrow's "well-digested conventionalism had incorporated itself thoroughly with his substance and transformed him into a work of art. Perhaps it was this peculiarity that invested him with a species of ghastliness and awe. It is the effect of anything completely and consummately artificial, in human shape, that the person impresses us as an unreality and as having hardly pith enough to cast a shadow on the floor." Although Allen may have feared his father because he seemed to see "straight through" him, his choice of pseudonym seems to hold his mother responsible for his anxieties about his identity and existence.

His sense that his mother threatened his intellectual identity also emerges in a dream of which he wrote, "I have had few other dreams that I can remember in such perfect detail." The dream concerns his Uncle Leo, his mother's brother, who "slept with his mother [Allen's grandmother] until he was twelve years old" and was later institutionalized.

> A few weeks after I had heard from a cousin that he was dead, I saw myself sitting on one end of a log and, on the other, Uncle Leo. We were in a thicket of scrub oaks. He was on the end of the log near the road, beyond which stood a four-story brick apartment building in which I lived on the top floor. There was no other building in sight. As I looked at Uncle Leo he fixed me with his large, brown opaque eyes, and I was afraid of him. To get to my apartment across the road I must somehow circumvent him, so I got up slowly, then began to run as fast as I could behind his back, dashed in, and got into the elevator, satisfied that I had outwitted a demon. He could not pursue me up the stairs as fast as the elevator could take me to my flat. At the top floor I opened the elevator door. There stood Uncle Leo barring my way.

Again, Allen professes not to know the meaning of the dream, while revealing it fairly evidently. "I have never understood how this poor man could play such an important part in my secret life. He was not only a gentleman, but gentle; ruined by a powerful, matriarchal mother."[12] Allen fears that his mother's presence, represented by Uncle Leo, even pervades his intellect, that "top-floor" apartment, and that all his conscious attempts to "outwit" her still cannot erase the effects of her subconscious influence; it rises as rapidly and inevitably to the surface of his mind as "demonic" Uncle Leo to the fourth floor.

After age nine, Allen found other intellectual influences, for when his brothers entered Vanderbilt University, his mother moved to Nashville with Allen and enrolled him in the Tarbox School. From 1909 to 1912, he attended the Cross School in Louisville, Kentucky. He began high school in Kentucky, at the public school in Ashland, where he remained from 1912 to 1914. His father's business ventures reunited the family for a half year in Evansville, Indiana, where Allen attended a local high school.

Like Caroline Gordon, he received part of his high school education in Ohio, when his father moved the family to Cincinnati in 1915, following another business failure. After a few months at the Walnut Hills School there, Allen enrolled at the Cincinnati Conservatory of Music in October 1916, where he studied until April 1917. Although the future poet had written some verse, he dreamed of becoming a great violinist. His biographer and friend, Radcliffe Squires, relates the fate of this ambition. "He had excellent teachers, studying violin under the Belgian violinist Jean Ten Have and the master Eugen Ysaye. His interest in violin remained with him into his later years, and one memory from the Cincinnati Conservatory enters importantly into his late poem "The Buried Lake." Even though many years later Paul Rosenfeld remarked on Tate's ability to play Mozart, the boy was not gifted. When Ysaye remarked that Allen was not creating music, it was a bitter blow."[13] He gave up his musical ambitions and decided to return to academic studies. After some work at the Georgetown University Prep School and tutoring in mathematics by the medievalist Dorothy Bethurum, Allen entered Vanderbilt University in the fall of 1918. Caroline Gordon, four years older than Allen, had already graduated from Bethany and was teaching high school.

Allen, like Caroline, was strong in the classics, but at Vanderbilt he developed an affinity for the study of English and American literature. Caroline belonged to a sorority at Bethany College; Allen joined a fraternity, Phi Alpha Delta. He also was a member of a literary club, the Calumet, and served as its president. Despite these parallels, Allen definitely had an advantage unavailable at Caroline's small Campbellite college—his extracurricular education in one of twentieth-century America's most stimulating and influential literary groups, the Fugitives.

The Fugitives began as a unnamed philosophical discussion group consisting of John Crowe Ransom, Alec B. Stevenson, Stanley Johnson, Walter Clyde Curry, Sidney Mttron Hirsch, and Donald Davidson. Their ranks were later swelled by Tate, Merrill Moore, James M. Frank, Jesse and Ridley Wills, and Robert Penn Warren. Although some of the members were affiliated with Vanderbilt as students or teachers, the Fugitive group, and later their magazine, had no official connection with the university.[14] Allen was introduced to the group by a young Vanderbilt English instructor, Donald Davidson. He was already acquainted with some of the others, notably, Curry, a medievalist who had helped Tate publish his first poems in the *American Poetry Review,* and Ransom, who had been his instructor in a composition course.[15]

As the members began to read and discuss each others' poetry, the group's focus shifted from the philosophical to the literary. From these critical discussions emerged the idea of publishing the best poems, and so a new little magazine was born. The magazine's initial manifesto reflected the youth of its editorial board. Although they were not sure what they were "for," they certainly knew what they wanted to rebel against: a Southern literary heritage that they perceived as a moonlight-and-magnolias nostalgia for the antebellum. "Official exception having been taken by the sovereign people to the mint julep, a literary phase known rather euphemistically as Southern Literature has been stopped up. . . . *THE FUGITIVE* flees from nothing faster than from the highcaste Brahmins of the Old South."[16]

With the conservatism of increasing years, many of them, including Allen, would recant; as the Agrarians of the early 1930s, they would even promote the values of a Southern heritage. Before they decided to take that stand, however, their youthful selves wanted to explore the new. As the editorial in the second issue ingenuously noted, "The group mind is evidently neither radical nor reactionary, but quite catholic, and perhaps excessively earnest, in literary dogma."[17]

In its relatively brief career from April 1923 to December 1925, *The Fugitive* gained an influence far beyond its size, place of origin, or members' reputations. After the first few issues, work by nonmembers was also published, including poems by older, established poets, such as the Mississippian William Alexander Percy, and new, unknown poets, like Laura Riding Gottschalk and Hart Crane.

Despite its literary success, the members discontinued publication with the issue of December 1925 because, in the words of the editorial, "there is no available Editor to take over the administrative duties incidental to the publication of a periodical of even such limited scope as the *FUGITIVE*. The Fugitives are busy people, for the most part enslaved to Mammon, their time used up in vulgar bread-and-butter occupations. Not one of them is in a position to offer himself on the altar of sacrifice." Despite this picture of aesthetes compelled to serve Mammon rather than Art, it seems likely that *The Fugitive* ceased publication simply because it had served its purpose of education and apprenticeship for the editorial board, many of whom, like Allen, were about to begin careers as professional men of letters.

During the Fugitive years, Allen had forged an identity as a poet. In many ways, his persona remained a version of the world-weary Henry Feathertop, born alienated from the present because of a head stuffed with the detritus of a more glorious past. However, his attitude toward his era was never entirely negative, as demonstrated in a short essay he wrote for the April 1924 issue of *The Fugitive*.

> An individualistic intellectualism is the mood of our age. There is no common-to-all truth; poetry has no longer back of it, ready for use momently, a mysterious *afflatus*—Heaven, Hell, Duty, Olympus, Immortality, as the providential array of "themes": the Modern poet of this generation has had no experience of these things, he has seen nothing even vaguely resembling them. He is grown so astute that he will be happy only in the obscure by-ways of his own perceptive processes. . . . So the Modern poet might tell you that his only possible themes are the manifold projections and tangents of his own perception.[18]

Using theatrical imagery prophetic of Wallace Stevens' 1940 "Of Modern Poetry," Allen resolves to make a virtue of necessity and be true to his own identity as a poet, however involuted. Many of Allen's critics would believe he dwelt too long "only in the obscure by-ways of his own perception," finding his poetry difficult and abstract.

Allen, however, did not believe that the poet's explorations of his consciousness entailed a withdrawal from the world. On the contrary, he was convinced that the poet's introspection answered a modern need, as in his definition of Fugitive: "quite simply a Poet: the Wanderer, or even the Wandering Jew, the Outcast, the man who carries the secret wisdom around the world,"[19] a goal the wandering Allen would certainly strive to achieve. Robert S. Dupree notes that the "poet's role, in the early poems, is to make his contemporaries acknowledge this harsh reality [a declining world]. He is responsible for making them uncomfortable, for dissolving with his acidic language their bland optimism."[20] Although one might perceive a certain arrogance here, as others sometimes did, Allen undoubtedly had a high sense of purpose as a poet and man of letters that he retained and acted on throughout his career.

As these formulations of the themes and purpose of modern poetry demonstrate, the twenty-four-year-old poet had developed into a formidable critic, both theoretical and practical. The Fugitives' method of meeting to criticize and defend each other's latest poems assisted Allen in developing his critical acumen, as John Crowe Ransom remembered. "In those many sessions he became confirmed for us as a critic whose quick unstudied judgment of a poem possessed authority. If it was protested, as was likely, he was able, by a process of thought so earnest and careful that it almost became visible, to make the unconscious grounds conscious; and as reasons they were valid."[21] Allen's "authority" as a critic would not only help him earn his living over the next decade, but would often serve as Caroline Gordon's touchstone for the quality of her fiction.

Through the Fugitives, Allen had acquired a literary family that he found much more satisfactory than his blood kin. Although he retained significant relationships with other Fugitives, his friendship with John Crowe Ransom and Robert Penn Warren are particularly important. In Ransom, more than a decade his senior, Allen found a father figure whose interests and judgments he could respect, if only as worthy of his opposition and rebellion. Curiously, Allen's memory of Ransom as his teacher is reminiscent of his account of his father's response to young Allen's filching of the whiskey.

> But I was not like him [Ransom], cold: I was *calidus juventa*, running over with violent feelings usually directed at my terrible family— terrible because my father had humiliated us. . . . My dislike of John was my fear of him. He had perfect self-control; I could see him flush with anger, but his language was always moderate and urbane. . . . His patience with my irregular behavior only made his disapproval all the more telling. What disturbed and challenged me most was my sense of the logical propriety of his attitude towards his students. He never rebuked us; his subtle withdrawal of attention was more powerful than reproach: he refused to be overtly aware of our lapses.[22]

Allen felt "fears" and "humiliation" before the "self-control" of those who refuse to be "overtly aware" of his existence, which he equates with his "irregular behavior." The fear before Ransom is the more intense because Ransom is refusing to acknowledge his *literary* "lapses" at the time Allen is establishing his identity as poet and critic.

The use of the word "lapses" and the portraits of cold and controlled father figures suggest that Tate's subconscious version of God the Father was a deity so indifferent to man that he would not even acknowledge his postlapsarian identity as a sinner. More telling, of course, is Allen's belief that his self's very existence is threatened without its sin and guilt. One might speculate that the duration of his marriage to Gordon, over thirty years, may be partially explained by her outspoken and often vituperative acknowledgement of his marital lapses. At any rate, Allen felt a constant need to rebel, to "disturb" and "challenge" himself through disturbing and challenging authority.

And "disturb" and "challenge" Ransom he did, by attempting to force his opinion of T. S. Eliot's high merit as a poet on the doubting Ransom. In 1923 Ransom published an article entitled "Waste Lands" for the book section of *The New York Evening Post.* Although he admired individual passages of *The Waste Land,* Ransom criticized Eliot's failure to provide a unifying framework for the fragments of modern life he presented. Allen responded with a letter to the editor attacking Ransom as a theoretical and practical critic, to which Ransom replied in a letter to the paper.[23] The exchange is telling in that it occurs in a public forum, even though Ransom and Allen are friends and correspondents. Ransom could use the public nature of the discourse as a means of retaining his self-control, and Allen could receive a public acknowledgement of his literary identity if he managed to arouse the ire of his mentor.

Allen's account of the incident in his memoirs is also significant. "I saw the attack [on Eliot] as the result of his irritation with my praise of Eliot, which was actually that of a distant disciple, to the neglect of him, my actual master from whom I learned more than I could even now describe and acknowledge."[24] In this version, written some fifty years after the event, Allen has reversed the way Ransom interpreted Allen's attack on him. In Ransom's letter of reply, he stated publicly that Tate was trying to liberate himself from the role of student and that Tate's letter was "but a proper token of his final emancipation, composed upon the occasion of his accession to the ripe age of twenty-three."[25] Although Ransom's appraisal seems just, he was once again refusing to acknowledge overtly Allen's lapse from literary gratitude and good manners. He veiled his anger with a "boys-will-be-boys" attitude that denied Allen's literary maturity. Predictably, Allen was infuriated and retreated from his friendship with Ransom for several months.

If through his relationship with Ransom, Allen was trying to rewrite the script of his dealings with his father, his friendship with Robert Penn Warren, on the contrary, provided him with a relation he had never had, a literary version of a younger brother. Warren, a Vanderbilt freshman and fellow Kentuckian, was more than five years younger than Allen. He met him in the spring of 1923 when Allen returned to the university after a withdrawal for illness. The respectful freshman at once asked him if he could see one of his poems, an immediate recognition of Allen's identity as a poet. They soon decided to room together with their fellow Fugitives Jesse and Ridley Wills.[26] Unlike Ransom, Warren shared Allen's opinion of T. S. Eliot's poetry, as Allen relates in his memoirs. "One day he applied art-gum to the dingy plaster [of their room] and when we came back we saw four murals, all scenes from *The Waste Land*."[27]

Allen not only gained from his association with the Fugitives, but also contributed a great deal. In addition to his critical acumen in their discussions, he provided them with a literary context that extended well beyond the confines of Nashville. Allen dated this more cosmopolitan consciousness to the spring of 1923, when he was absent from the university to recuperate from an illness.

> In the nine months of my absence from Nashville I think I began seriously to study the writing of poetry, and I began to be a little more aware of the world, or at any rate of the literary world, at large. In May, Hart Crane had seen one of my poems in *The Double Dealer,* a 'little' *Dial* published in New Orleans. He wrote me a letter from Cleveland and sent me some back numbers of *The Little Review.* He said that my poem showed that I had read Eliot—which I had not done; but I soon did; and my difficulties were enormously increased. Anyhow, from Eliot I went on to the other moderns, and I began to connect with the modern world that I had learned about from Baudelaire, first through Arthur Symons, then from Baudelaire himself. I mention this personal history because I believe it was through me that modern poetry made its first impact on the doctors [The Fugitives] who gathered fortnightly at Mr. Frank's house.[28]

As Allen introduced the Fugitives to modern poetry, he was also preparing himself to enter the scene of literary modernism.

He graduated from Vanderbilt in the summer of 1923, served briefly as assistant editor of *The Fugitive,* and then obtained a position as a substitute teacher in a Lumberport, West Virginia, high school where he remained from March to June 1924. At the end of the school year, he was finally able to make his long-anticipated trip to New York, which he regarded as the center of his literary universe, a feeling shared by many of the younger poets and writers already on the scene.

Malcolm Cowley, then a young poet from Pennsylvania, remembered reading and discussing Poe with Allen and Hart Crane in Crane's apartment with its view of the Brooklyn Bridge, the view that inspired Crane's American epic, *The Bridge.* The three young writers went out to continue their conversation at the end of the pier. "Suddenly," Cowley wrote, "we all felt—I think we all felt—that we were secretly comrades in the same endeavor: to present this new scene in poems that would reveal not only its astonishing face but the lasting realities behind it. We did not take an oath of comradeship, but what happened later made me suspect that something vaguely like this was in our minds."[29]

Allen remembered himself at this time as something of a dandy, carrying a cane and wearing his Phi Beta Kappa key.[30] Cowley recalled a trait that remained characteristic of Allen, his detached courtesy: "I thought he used politeness not only as a defense but sometimes as an aggressive weapon against strangers."[31] This elegant, courteous, and somewhat arrogant young modernist was unemployed, though, and had to leave New York and his new friends and return to the South and an old friend. His fellow Fugitive Robert Penn Warren had invited him for a visit at his parents' home in Guthrie, Kentucky.

Guthrie was and is a small town near the Tennessee border north of Clarksville. In a letter, Allen described it as quite rural, yet in the summer of 1924 it contained three fledgling writers who would rise to prominence with the Southern Renaissance.[32] Tate and Warren were joined by Caroline Gordon who was on vacation from her job as a Chattanooga reporter and visiting her parents at the current scene of her father's ministry. Warren was a somewhat shy red-headed nineteen year old, almost a decade younger than Caroline. Although she had not met him before, "Red" may have had some of the aura of the boy next door since his parents and hers were acquainted and part of the web of Guthrie's society.

Warren's friend Tate, newly arrived from the New York avant garde, must have been an exotic presence in Guthrie and a breath of fresh air to Caroline Gordon. At twenty-three, he was almost five years older than Warren, but his air of worldly courtesy made him seem older, more of a contemporary of twenty-eight-year-old Caroline than of his friend Warren. She already knew him by reputation since she had written admiringly of the Fugitives for the Chattanooga *News*. He shared her literary aspirations and her desire to leave the South for the artistic Mecca of New York. Allen was handsome, charming, and well into his career as a formidable ladies' man. Caroline had a distinctive, unconventional kind of good looks, dark and intense with a hint of melancholy. She was an "older" woman, but deferred to his opinions in literary matters, considering herself a somewhat provincial writer of unpublished fiction. In the limited world of Guthrie, it is not surprising that a passionate attachment ensued.

The official story here, the one Tate and Gordon used for public consumption, such as their *Who's Who* entries, is that this romance had a conventional happy ending, leading to their marriage that fall, on November 3, 1924, in New York City, and the birth of their daughter Nancy on September 23, 1925. Unfortunately, the facts contradict this account but explain a great deal about the later history of the Tates' marriage.

Allen returned to New York in October. Through a friend of Malcolm Cowley, Susan Jenkins, he got a job as assistant editor of a pulp magazine, *Telling Tales,* which Jenkins edited for the wonderfully named Climax Publishing Company. His job was undemanding, mainly copyreading, and left him plenty of time for his writing and an active social life.[33] During his New York years, he promoted his literary reputation, and made some extra money, by reviewing for *The New Republic, The Nation, The Bookman, The Hound and Horn,* and *The Sewanee Review.* Many of those he reviewed were also aspiring poets who were, or would become, his friends, such as Phelps Putnam, Hart Crane, e.e. cummings, Archibald MacLeish, Malcolm Cowley, Leonie Adams, Louise Bogan, John Peale Bishop, Mark Van Doren, and Yvor Winters.[34]

In the evening, Allen might attend a burlesque show at the Winter Garden with Hart Crane or stop for a hot rum toddy at a speakeasy known as the Poncino Palace. A favored gathering place was Squarcialupi's restaurant on Perry Street. According to Malcolm Cowley, Squarcialupi's habitues included "Kenneth Burke, Allen Tate, Hart Crane, Matthew Josephson, John Brooks Wheelwright, and William Slater Brown, then husband of Sue Jenkins Brown, besides Sue herself, Peggy Cowley, and Hannah Josephson, with an occasional visitor from Paris and a girlfriend of one of the bachelors." They read their poems aloud, encouraged the proprietor to sing, and put together a compendium of their work, *Aesthete 1925,* as they drank red wine and ate dinner.[35]

Caroline Gordon is not mentioned as among those present in various accounts of that winter of 1924–25. In 1968, Allen Tate wrote to Cowley about a piece Cowley had written for the *Sewanee Review.* To avoid reinflaming Caroline's feelings, he asked him to delete a reference to Caroline's absence from their parties during that winter.[36]

If she never appeared at their parties, she was leading a parallel life of work and activity of her own. Before she went to New York that fall she spent some time with her cousin "Little May" Meriwether, now Mrs. Sherman Morse, at Canandaigua Lake near

Rochester, New York. There she met a young painter, a native of Rochester, who gave her his sister's address in New York. Her name was Sally Wood, and she was a nurse who had served in the First World War and was currently married to a labor organizer, Stephen Raushenbush. She and her husband lived in Greenwich Village, and it was to their apartment that Caroline came on her arrival in New York. Sally Wood wrote:

> I was alone in our apartment one day when the doorbell rang. There stood a memorable figure, a young girl like and unlike me. We were about the same size, but her coloring was startlingly different. My hair was a nondescript brown, while hers gleamed blacker than a raven's wing, matching her eyes set in a masklike face, magnolia white. . . . She said at once, or at least very soon, that one was never so happy as when asleep, and I must have shown her a couch, because she dropped off in a minute.[37]

The two women became fast friends, but as Miss Wood's recollections indicate, Caroline Gordon was always a romantic, devil-may-care alter ego to her friend. Caroline continued to play this role in her letters to Sally Wood, which generally transform bitterness, pain, and uncertainty into a marvelous string of comic anecdotes.

Gordon did not remain on the Raushenbush's couch, however, but roomed with a newspaperwoman whom she had known in Chatanooga, Vivian Brown known for her trademark big, floppy hats. Caroline found intermittent newspaper work, including some legwork for the syndicated columnist Berton Braley, but she was backed financially by her father.[38]

In letters from her parents, written in the spring of 1925, there is no mention of Allen and her parents assume she is rooming with two other young women, although Allen and Caroline were married on May 15, not the previous November.[39] Allen is first mentioned in a letter from her father on July 2, 1925, and on July 13 her mother writes, "I am addressing this letter to Carolyn Gordon. I see our letters have been going to the dead letter office because they were addressed to Mrs. Allen Tate & 'there ain't any.' " By August 1925, she is receiving letters at 30 Jones Street as Mrs. Allen Tate, and the Tates' daughter Nancy was born on September 23, 1925.

Toward the end of his life, Allen Tate told Nancy that although

Caroline was pregnant, she at first refused to marry him because she had seen him with a beautiful woman in a restaurant as she walked by outside. She was very jealous and very proud.[40] Laura Riding, a poet published in *The Fugitive,* believes herself to have been a further romantic complication. "Tate and I met in 1924, he then pronouncing me the woman to save American poetry from the Edna St. Vincent Millays. There followed a correspondence, during which he wrote with fervor as one most seriously committed to me personally: but a preceding commitment to Caroline Gordon took over."[41] Curiously enough, Laura Riding seems to have had some of Caroline Gordon's characteristics, such as a vigorous intellectual tenacity that some might regard as obstinacy. After visiting her in Louisville in February of 1924, Allen wrote of Riding, "Her intelligence is pervasive. It is in every inflection of her voice, every gesture. . . . But always you get the conviction that the Devil and all Pandemonium couldn't dissuade her of her tendency."[42] "Laura Riding Roughshod" was Hart Crane's appellation for his fellow poet.[43]

Malcolm Cowley remembers hearing a different version of events.

Allen was a ladies' man. He was very attractive to ladies, and he had terrifically good manners. That was really his offensive weapon. Well, one of them, he confided to Sue [Jenkins], became pregnant. It took her a long time to find out she was pregnant. . . . She had symptoms of discomfort and went to various doctors, two or three anyway. Finally she went to an osteopath. The osteopath told her, 'Young woman, you're pregnant.' So it was thought that Allen and Caroline should get married so that the baby would be legitimate, but Allen didn't want to get married at all. The agreement was that they would get married and separate immediately.[44]

In all versions, Allen is characterized as a "ladies man," and the marriage as reluctant, but with the reluctant partner varying. It would be understandable if both impecunious young writers felt doubts about beginning married life under these circumstances, but the Tates were united by more than a pregnancy and a marriage ceremony, for they did not separate after their daughter's birth, but remained together for over thirty years. Each met certain of the other's needs and the pattern of their relationship was established at this time.

This pattern is delineated in one of Allen's poems, published in the December 1924 issue of *The Fugitive,* entitled "Fair Lady and False Knight."[45] The lady has been seduced and abandoned by the knight in "the twilight fields of June / Where once a chit I played," which may be Allen's imaginative use of his relationship with Gordon in June of 1924 in the territory of her childhood, along the Kentucky-Tennessee border. The lady does not submit to this treatment with resignation but curses the fields and landscape with vigor, as Caroline did Allen's transgressions. She is described as very white of face and "strict and crimped" of mouth, both of which could be said of Caroline Gordon, although the lady of this poem has come into this state as a result of her abandonment. The knight is curiously passive and unsympathetic, as in this stanza that concludes the poem.

> Alas, Fair Lady, for death shall be,
> I trafficked your lips like gold,
> And I was the one sweet headlong star
> Your eyes will ever behold.

His primary interest is his apotheosis as the lady's only passion. Since her imminent death will allow him to retain that exclusive status, his "Alas" over her demise seems quite perfunctory.

Whether this particular poem has its genesis in Allen's relation with Caroline is, of course, speculation, but it clearly delineates his self-image in relation to women. The knight believes in loving her and leaving her. He wants to believe that he has left her with a precious memory, but knows that he has grievously wounded her. This particular pattern suits a romantic relation with a woman, but it fits into a larger pattern in Allen's dealings with others. Like Allen in his relation to his father and Ransom, the knight seems to need to transgress in order to provoke his victim into acknowledging his existence, albeit with curses.

The fantasy that transfixed Caroline's imagination interlocks almost too neatly with that of Allen. Her fiction is filled with women who are abandoned when their men heed the siren call of another obsession. In her Civil War fiction, it is the summons to battle. In her novel and tales about Aleck Maury, based on her father's recollections, it is the lure of sport. In her later fiction, the siren is indeed a siren, and the issue of adultery is openly addressed. This drama of betrayal and abandonment may stem from her father's evasions of her mother through sport, but it may simply be Caroline's view of the woman's role in a man's world: secondary, and easily shunted aside.

Although the Tates' mutually reinforcing, or destructive, fantasies are confirmed in their writing and their behavior, a marriage that lasted more than thirty years indicates a firmer basis than symbiotic failings. As they confronted life together, they were also united by mutual love, a strong physical attraction, a delight in each other's intellect and wit, and the fostering of each other's strengths. Despite their troubles, each gained from the marriage in ways with literary as well as personal ramifications.

Their immediate problem, however, was financial: how to pay for Caroline's prenatal care, the hospital for the birth, and the food and environment an infant needed. Malcolm Cowley wrote an appeal for five hundred dollars to the Personal Service Fund on Tate's behalf, characterizing his situation as unusually difficult since he was making only thirty dollars a week on hack work. Cowley also told the Fund that Caroline needed to see a doctor twice a week and the couple had no savings.[46] As Caroline recalled, they soon lost even their weekly income from *Telling Tales*. "One day the owner sent out a note to the effect that . . . he expected all the editors to turn in copy [that] was grammatically correct. Unfortunately, there was a grammatical error in the note. Allen took it upon himself to correct it. He was fired, with a 'severance pay' of forty dollars."[47]

Somehow, they did manage, and Nancy was born on September 23, 1925, at Sloane's Lying-In Hospital. In attendance was Caroline's friend Sally Wood, who later wrote: "At the end I was sitting by her bed at the Sloane Maternity Hospital, timing her labor pains, and Caroline was grimly talking about her novel, her white face flinching now and then. A door opened, and thinking it was a nurse, I turned around. What was my surprise to see Allen, clutching his cane (he never dressed like a Villager), his face more ravaged than Caroline's." Apparently he was suffering the ravages of alcohol as well as imminent paternity since, as Caroline told Sally Wood, "his friends had clustered round him . . . before she went into the hospital, determined to see *him* through it, plying him with what passed for liquor in those days."[48] Wood characterizes Allen as a frivolous observer and Caroline as a dedicated artist, or perhaps one who desperately clings to her dream of becoming a writer and so needs to distract herself from the threats to her time and concentration an infant would bring.

According to Laura Riding, she accompanied the family of three home from the hospital. She recalled, "it chanced that I lived not far from the Tates in Greenwich Village. We were friendly neighbors. I carried the baby when he brought Caroline home from the hospital by the 'elevated.' I left at years' end for abroad. There was no further contact between us."[49] From this account and Gordon's remark about Allen's friends during her confinement, she must still have felt very much the outsider, connected to the Village intellectuals only through Allen, not yet as a personality or writer in her own right.

The baby was named in accordance with Meriwether family tradition. Caroline wrote, "Allen with the typical imperceptiveness of a youthful father, planned to name her for his mother [Eleanor] and his 'boss' on the magazine, Susan Jenkins. I had to remind him that our family had a chain of Nancys and Carolines alternating for generations. . . . Being a Virginian himself [by descent], Allen realized the power of tradition and gave way in the matter of our daughter's name."[50] Caroline may have been asserting her own identity and her "group" in the face of Allen and his circle of friends. She was, however, also setting up a Meriwether-Gordon claim to Nancy that would soon assert itself.

The Tates moved to a larger apartment at 47 Morton Street, but they apparently had trouble filling some of that greater space, in particular the cupboard. In answer to an appeal from Caroline, her mother arrived from Kentucky only to find that there was no food in the house.[51] With her customary vigor, Mrs. Gordon took charge of the grocery shopping and Nancy. On October 20, Allen wrote to Cowley, with some irony, that since her mother's arrival, Caroline was a bit more at liberty—from the baby. Caroline may have had more free time, but she was still under her mother's domination.

Nancy Gordon wanted more permanent control over her namesake. She planned to take her home to Kentucky, thereby repeating the frequent attempts of her own mother, Miss Carrie, to keep her granddaughter and namesake, Caroline Gordon, with her.

Caroline consented to her mother's plan, later explaining, "I was feeling so feeble I couldn't combat the forces that were operating against me." These forces were threefold, according to Caroline. She dreaded her mother's pain if she could not have Nancy. "I saw mother would have collapsed. She had born it all amicably—wild people . . . dropping in—just so she could get Nancy." The second force was her fear of imposing an additional burden on Allen. "I was afraid Allen would break down trying to work day and night too."[52] Finally, she was concerned for Nancy's welfare: "I tried to console myself by reflecting that if Nancy had stayed with us she might have starved to death."[53]

The decision to give up Nancy until their finances improved only reinforced Caroline's postpartum depression. She wrote to Sally Wood, "I've felt paralyzed for weeks, You see, I only really had her one day for myself."[54] She suggested that she had served everyone else's interests but her own, but her depression may have been deepened by suspicion about her true motives. She needed to keep her new husband and she needed to gain some freedom for her writing. Hackwork office jobs for Allen while she cared for Nancy would have furthered neither goal, but she could have kept Nancy. No solution would have been completely satisfactory to all involved, and so guilt was unavoidable.

The Tates still needed to find a quieter and cheaper way of life, and a visit they had made the previous summer suggested a solution. Allen's former boss, Susan Jenkins Brown, and her husband William Slater Brown had purchased a pre-Revolutionary farmhouse in the vicinity of Pawling, New York. The house was near a cave that was "a refuge for a band that preyed on the two-wheeled carts supplying Washington's army encamped in the area"; it was called Robber Rocks by Patriots and Tory Hill by Loyalists. With the help of Hart Crane, the Browns restored the house and gave a memorable housewarming party to celebrate the Fourth of July in 1925.[55] The pregnant Caroline attended with Allen. Hart Crane described the party in a letter to his mother:

> Nothing could beat the hilarity of this place—with about an omnibus-full of people here from New York and a case of gin, to say nothing of jugs of marvelous hard cider from a neighboring farm. You should have seen the dances I did—one all painted up like an African cannibal. My makeup was lurid enough. A small keg on my head and a pair of cerise

drawers on my legs! We went swimming at midnight, climbed trees, played blind man's bluff, rode in wheelbarrows and gratified every caprice for three days until everyone was good an' tired out.[56]

Despite this seemingly ceaseless gaiety, Susan Jenkins Brown remembers that on this visit Caroline and Allen found a nearby house for sale, and "wistfully looked it over, though, as Caroline said, 'We couldn't even buy an extra Woolworth dinner plate' "[57]

Since Allen was no longer employed, they were no more solvent in November than they had been that July, but Caroline was no longer pregnant and the infant Nancy was in Kentucky, so the Tates decided to try spartan rural living. If they still could not afford to buy, they could pay rent, which was considerably cheaper than in the city. Half a mile away from the Browns, Mrs. Addie Turner shared a large farmhouse with her aunt. The Tates could rent eight rooms of it for ten dollars a month, but they would have to share a communal pump in the kitchen and provide their own heat by chopping wood for their stoves. Susan Jenkins Brown's account of their arrival seems to foreshadow the Tate's year: "They moved in on a miserable, rainy November day, with Bill [Brown] driving them from the station in our $35 Model T, barely able to make it through the muddy, rutted dirt road."[58] Their experience in the Tory Valley was a struggle, but they succeeded, in terms of their survival and their marriage.

As if the rigors of the winter were not enough, the Tates decided to invite Hart Crane to join them. They felt that without the metropolitan distractions of alcohol and sailors, Crane might get some work done on *The Bridge.* "I never heard of such feckless generosity," said Malcolm Cowley.[59] One can only agree since a housemate known for his wild spontaneity and need for distractions could hardly be the best choice for a couple bent on solitary supplication of the muse. It also seems curious that a couple beginning married life would want a third party, but actually the invitation to Crane established a pattern in the Tates' married life. They often had others living with them, those who needed help, whether financially, emotionally, or artistically. The third party acted as a buffer between them and ultimately as the lightning rod that would receive the blast of their mutual tensions, the blame for the failure of the extended household. And so it was with Crane.

The experiment began badly because the Tates did not find

themselves in their anticipated position of patrons to a penniless poet. According to Crane, Caroline later told him that "as soon as they found out that I had been fortunate in acquiring funds they immediately began to doubt the advisability of inviting me out."[60] Crane had received a grant of one thousand dollars from Otto Kahn, a wealthy sponsor. According to Tate's biographer, Radcliffe Squires, "Hart finally reached the Turner house on Saturday, December 12, after several nights celebration. Instead of being despondent and empty-handed, he arrived with liquor, fancy groceries, a new pair of snowshoes, and extravagant plans of work for the coming year."[61] Roles were reversed, and the Tates' protégé actually lent them money.

Crane could be a man of magnetic charm, whose unpredictability was a source of fascination as well as disturbance. Susan Jenkins Brown points out that "Tantrums and violence can so readily be recalled that they tend to blur recollections of more admirable but less dramatic elements. One can recall the sensation of spontaneous joy and laughter, but not often can one reproduce the wit and humor that provoked it."[62] Over forty years later, Caroline Gordon wrote of Crane, "Foreign travel was immensely stimulating to him, but he brought the same intensity to everyday life. A walk along a country road, a croquet game in the rain . . . his imagination transformed these homely incidents into memorable events. It was such infectious high spirits, plus his genuine concern for their welfare, which extended to their children, that so endeared him to certain of his friends."[63] Characteristically, Crane began his sojourn in the Tory Valley with an extended holiday party. He wrote to friends on New Year's Eve 1925, "Nothing much *yet,* however has been done by any of us. There have been food supplies, appliances, and sundries of all sorts to order—and then came Christmas with flocks of people visiting Brown's place . . . and drinking and talking day and night."[64]

After the holidays, the three writers established more of a routine and set to work. Crane did the dishes, Tate made breakfast, and Gordon made lunch and dinner.[65] Crane wrote to his mother on January 7 that "I've been at work in almost ecstatic mood for the last two days on my *Bridge.* I never felt so much range and symphonic power before."[66] Allen appears to have been in a state of poetic gestation, and spent much time roving the hills with his new gun, "The White Powder Wonder," occasionally bagging some small game. Caroline remained within, attempting to begin her second novel.[67]

By the end of March, however, cabin fever had set in. The winter had been quite severe, and the three writers were often restricted to the premises with no outlet for their energies aside from cutting wood to keep themselves warm. Distractions were further reduced, and feelings of confinement increased, when the weather became so bad that their mail could not be delivered for weeks. Crane became tired of what he called a "life of perfect virtue, redundant health, etc.," and the Tates appear to have become tired of Crane.[68] When his inspiration gave out, he sought their company, but what was comforting diversion to him was annoying distraction to them. Doors began to be locked between parts of the house, and at one point the two camps communicated only through their landlady, Mrs. Turner, or by notes slipped under the door.[69]

To the Tates' withdrawal Crane responded with amazement, but he blamed the actual quarrel on Caroline. He wrote to his mother on April 18 that "primarily it has been Mrs. Tate who has influenced matters until they came to a head the other day." According to Crane, Caroline was the one who put the bolts on the doors and asserted that "I had from the moment of arrival proceeded to spread myself and my possessions all over the house, invading every corner." Crane adds, however, "The contents of Mr. Tate's letter, were about the same, a little more gracefully phrased that's all." Crane assures his mother that "I'm not quoting *insinuations* in any of this, I'm using practically their own words."[70]

If Crane is paraphrasing closely and accurately, Caroline may have been demonstrating her territorial rights over Allen to the homosexual Crane since her words and actions concern the warding off of an invader. On the other hand, since Crane states that Allen echoes her sentiments, Crane may simply have preferred to blame a woman, rather than a man and a fellow poet. Allen seems somewhat passive, the warred over, rather than the warrior.

Crane made a temporary retreat to a mother figure. He began to take his meals with Mrs. Turner, who would exclaim with sympathetic wonder, "Mr. Crane's *so* sensitive and nervous."[71] Unable to withstand what he called "the hypnosis of evil and jealousy in the air," he left on April 28 for the Isle of Pines.[72] His quarrel with the Tates was not permanent, however. At the last minute, Allen would write the introduction to Crane's collection *White Buildings* when Eugene O'Neill could not honor his commitment to do so. Several months after Crane left, Caroline wrote to Sally Wood that "Hart is a very fine poet, but God save me from ever having another romantic in the house with me!"[73] In later years she spoke of him with amusement and affection.

After their quarrel with Crane, the Tates seemed to find new satisfactions and pleasures in each other. In March, when Caroline made a visit to one of her brothers in Washington, D.C., Allen complained to Cowley of his solitude and loneliness, adding that if Caroline did not return soon, he would undoubtedly get scurvy from his diet of cornmeal and rice, all that he could manage to cook by himself.[74] To Caroline herself, he wrote more of his need for spiritual sustenance since he found himself unable to settle down to work without her. He hoped that she would not use his weakness for her own ends, but spoil him a bit.[75] Caroline's role as Allen's catalyst, and her drive for power, are as evident here as Allen's new dependency on Caroline. He is now eager, not reluctant, for a spouse. Caroline also found Allen a much improved spouse, writing in the fall of 1926 that he "has changed a lot in the past year. He's certainly a more integrated person."[76]

The responsibilities of parenthood were also to be integrated into their new identity as a couple, however briefly. At the end of the summer, Mrs. Gordon brought Nancy, now almost a year old, on a visit to her parents. Her weekly letters, and those of her husband, were detailed bulletins concerning every aspect of Nancy's health and development.[77] Nancy's grandparents clearly doted on her, but the modern Tates were skeptical of Mrs. Gordon's childrearing methods. Caroline wrote to Sally Wood:

> It is no worse than I foresaw when I let Nancy go—in fact I painted the picture for Allen at the time. He says he thought that I was exaggerating, and that my fears were due to my nervous condition—but now admits that it is all exactly as I prophesied. . . . She is medieval in spirit. . . . In the rearing of Nancy she hopes to correct all the errors she made with me! Poor Allen is frequently quite embarrassed when she says to him quite naively, "and you see how Carolyn turned out." He realizes that he is the bad end to which I have come, in Mother's opinion. . . .[78]

Once again, the Tates closed ranks against a third party. "My disapproval of Mother, my indignation against her—and Allen's— is all on moral grounds. And she feels great disapproval of us too, so we're completely antagonistic."[79]

Despite these strong reservations, the Tates once again allowed Mrs. Gordon to take Nancy back to Kentucky. Caroline's reasons were much the same, citing the needs of three people other than herself. Nancy needed regular nourishment and warmth that the Tates' "uncertain fortunes" could not guarantee. Again, Caroline feared that her mother would "collapse" if they took Nancy from her. The third person to be propitiated, though, is no longer Allen but her father: her mother "always makes him suffer for any disappointment that comes to her. He and I have had a gentleman's agreement ever since I was fifteen or sixteen to help each other out when we can."[80] Allen's interests are now presumably one with those of Caroline.

Adjustments and compromises had been made to arrive at this state of harmony. Caroline loved country life. She wrote to Sally Wood "I'm not at all an urban person, you see. I love to have space around me, and I love to dig in the dirt and walk in the woods." Unfortunately, after his initial enthusiasm, Allen's interest in hands-on gardening waned, but he and Caroline worked out a compromise: if he would do the dishes, she would hoe the garden. As Allen lost interest in the practice of farming, its intellectual and cultural implications increasingly preoccupied him, a change Caroline somewhat satirically noted: "He has the strangest attitude toward the country—the same appreciation you'd have for a good set in the theatre. I think Allen feels toward Nature as I do toward mathematics—respectful indifference. He walks over the garden hailing each tomato and melon with amazement—and never sees any connection between planting seeds and eating fruit."[81] Caroline portrays Allen as the pastoral poet true to the tradition of the genre, highly conventional and not at all earthly.

Allen may have been respectful toward nature, but he was not indifferent. Like many of his fellow Fugitives, and Southerners in general, he was reevaluating his attitude toward his rural Southern heritage in the wake of the 1924 Scopes trial. The controversy over the teaching of evolution in Tennessee schools caused Southerners to be portrayed in the national press as ignorant and backward hicks; the South had seemingly confirmed H. L. Mencken's epithet, "the Sahara of the Bozarts." By the summer of 1926, Allen was engaging in the correspondence with Ransom and others that was the seed of the Agrarian movement. Since, in the Agrarians' view, the North was irredeemably industrialized, the self-sufficient Southern yeomen farmer became their ideal.

During that summer of 1926, Allen was not, however, reversing his earlier espousal of modernism as he began to meditate on his Southern heritage and his great "Ode to the Confederate Dead."[82] This poem announces the way Allen's writings would attempt to reconcile the doubt-filled and fragmented modern mind with what to him seemed the wholehearted and unified beliefs of the past. In this poem, Allen's persona, a modern man, considers the way the beliefs of the Confederate soldiers led directly to their actions, which, in turn led to their death and burial in the graveyard he contemplates. The speaker does not know how to respond to their action nor how to act himself. He asks,

> What shall we say who have knowledge
> Carried to the heart? Shall we take the act
> To the grave? Shall we, more hopeful, set up the grave
> In the house? The ravenous grave?

Allen's Agrarianism and his espousal of Southern values was never simpleminded or simplistic. He recognized the burdens as well as the benefits of the past, and he realized that he lived in another age. His greatest work is evoked by his attempts to answer the questions about the relation of past and present that he asks in the "Ode to the Confederate Dead"; his ability to hold these irreconcilable concepts in tension provided him with a poetically fruitful, if paradoxical, unity.

Although Allen is responding to national issues, Caroline's role in his new interest in the South and agrarianism is significant. She sang the praises of rural life into the ears of this former hotel child and helped him leave the metropolis he so loved. A certain pride in her Southern background emerged in the stories she told him about the rural Meriwether connection, stories she would rework into her fiction over the next decade.

At this time, however, Caroline herself had not discovered these themes for her own work. Her first novel was the work of autobiographical fiction called *Darkling I Listen,* which she had written in her aunt's attic in Chattanooga. Its fate makes the importance of Allen's influence on Caroline quite clear. She gave it to him, and watched his face as he read it. She said that she could tell by his expression "that it wasn't any good," so she destroyed it.[83] Whether Allen was right or wrong or whether Caroline even interpreted his facial expression correctly, we will never know, but her respect for his judgment is striking.

During the Tates' year in the Tory Valley, from November of 1925 to the fall of 1926, Caroline was sporadically at work on her second novel, but does not seem to have made much progress. In November, she appears to have been too busy settling in to write much. With December came holiday festivities and Hart Crane. Her lack of progress in January may be explained by her literal and figurative position in the household: she wrote in the kitchen, and, until Crane lent her a spare typewriter, she had to wait until Allen's was not in use.[84]

After a brief spurt of work in February, Caroline was distracted by the quarrel with Hart Crane and her trip to Washington. She then took over Crane's former study and comments, "I find that having a room of my own enables me to write—I couldn't write a word all winter."[85] Unfortunately, she seems to have been side-tracked once again for in a letter to Sally Wood, she mentions writing "potboilers" to earn the trainfare for her mother and Nancy's visit.[86] In another letter to Wood, dated September 9, 1926, she remarks that the outline of her second novel is now in the hands of an editor.[87] Presumably, it was not accepted since her first novel printed was *Penhally,* brought out in 1931 by Scribner's, and Malcolm Cowley believes the novel on which she was working during this period was never published.[88] Neither the novel nor the outline survive in her papers.

The outline is an accomplishment, since she produced one she was willing to expose to a publisher's eye, but it is also an acknowledgement of failure since the novel itself is not yet written. She was undoubtedly experiencing the stress of adjusting to marriage and the guilt of long-distance motherhood, but, to a certain extent, she allowed herself to become more distracted by adding someone like Crane to her household, working in the kitchen when there were eight rooms from which to choose, and dropping her work to meet the demands of others. She simply did not have the confidence in herself as a writer to take seriously her needs for time and space.

The fault, of course, does not entirely belong to Caroline Gordon since the mores of the day, even among the Greenwich Villagers, would not buttress her sense of purpose. In a 1984 interview, Malcolm Cowley confessed, "You have to get the admission of an aged fellow that I was a little bit antifeminist at that time. That is, in our discussions we were the boys. The boys always got together, and the girls weren't asked to join them."[89] By "the girls" Cowley meant Caroline and two other fiction writers, Josephine Herbst and Katherine Anne Porter. Only on the Tates return to New York in the fall of 1926 did Caroline find in these women a "support group" of her own. With their encouragement and that of an older male novelist, Ford Madox Ford, Gordon would make the leap in commitment from unpublished amateur with work perpetually in progress to a published professional.

Before this happy day arrived, however, the Tates needed to confront the writer's chronic problem of making a living in a way that still left enough time and energy to write. They knew they could not face another winter in the country, and so decided to return to New York where they obtained a basement apartment at 27 Bank Street in return for Allen's services as janitor. The nature of his employment became a touchy issue when Allen's friend, the journalist and critic Matthew Josephson, wanted to write a sob story about the poet forced to work as a janitor. Tate recalled in 1962 that he would have no part of that kind of sleazy exploitation.[90] According to Malcolm Cowley, Caroline saved the day: "Having been a newspaperwoman herself, instead of saying, 'Oh, no, don't write about it,' she expressed great eagerness to have her name in print, and that's what killed the story. If you want not to be written about, just express great eagerness."[91]

Caroline, too, took a job, but hers was a source of excitement, not embarrassment. The English novelist Ford Madox Ford wanted a secretary to help him during his sojourns in New York and look after his mail when he was away. Caroline got the job, and began her lengthy connection with one of the twentieth century's most curious characters. He was large, fat, and fair with an unattractive habit of breathing heavily through his mouth. According to Allen, "Ford's best biographer will understand at the outset that Ford himself must be approached as a character in a novel and that novel a novel by Ford."[92] Like Henry James, whom he regarded as the Master, Ford lived for his art, and he made his life into art by fictionalizing his memories and anecdotes, in works which were ostensibly nonfiction, such as his autobiographies.

Ford was also known for his tact and generosity toward struggling young writers. Allen recalled that "One day he brought me a sentence, and like a beginner he asked me, a beginner: 'Do you think it will do?' He could ask this because the dignity and unremitting demands of his art came first."[93] Ford was also building Allen's confidence by showing him that the hard search for *le mot juste* was not exclusive to the novice, but common to every professional writer. Caroline remembered that Ford kept a corner of his desk piled high with the work of young writers: "One morning I laid a manuscript of poems by Leonie Adams on top of the heap. He read the poems . . . and folded his hands on top of the heap and said, 'Tell her that I am completely at her service.' "[94] He honored young writers as fellow craftsmen; by treating them as professionals he helped make them into professionals.

Although Caroline was in close contact with Ford and saw him encourage young writers, she did not show him her work until a couple of years later in Paris. Malcolm Cowley and others recall Caroline writing at this time, but she did not share her work with others. She herself said that "I never have anything in a state to show anybody," the kind of perfectionism that makes obtaining constructive criticism, or completing a work, quite difficult.[95]

Even before Ford started helping her with her manuscripts, she learned some lessons from his professional attitude toward his work and from the techniques he exhibited in his fiction. Caroline called Ford her "Master" and "the best craftsmen of his day," adding that "There is no one, not even James, who can bring a scene before us with more vividness."[96] The scene may be vivid, but the reader remains outside, as Caroline noted when she described the "surface" of Ford's works as "luminous and impenetrable," an epithet that could apply equally well to Gordon's fiction.[97] For Caroline, though, Ford's greatest technical achievement was his manipulation of time shifts, which she sought to emulate in her own work, particularly *Penhally*.

Her affinity with him extended beyond matters of technique,

into characters and themes. She identified La Belle Dame Sans Merci as the key figure in Ford's works. This woman who seduces, and abandons, without remorse is the female version of the masculine betrayer found in Caroline's own work, demonstrating the similarity of their world view.[98]

In an article published in 1954, "The Story of Ford Madox Ford," Caroline evaluated the shape of Ford's career as a whole. She praises him as a historical novelist at whose "touch some of history's driest, barest bones take on flesh, but laments that "he is comparatively unknown as a historical novelist in an age in which the historical novel enjoys the greatest vogue it has ever enjoyed." She then considers the reasons for this unwarranted neglect.

> It has been fashionable to regard this obscurity, deeper and darker than that surrounding any comparable talent of our time, as no more than he deserved: the proper reward of a misspent life. Ford during his lifetime was often the subject of gossip, his actions seemed ill-advised; he made powerful enemies. In his late fifties his powers failed him; he was no longer able to write fiction and kept himself going by writing over and over a sort of fictionalized autobiography.[99]

All these comments could apply to Caroline's career as well. Her historical fiction never sold particularly well; she made enemies of critics who could have furthered her reputation; by her late fifties, she too was writing a fictionalized autobiography she never completed, and some critics believe her powers had failed her. Although in 1954 Gordon is projecting her current state onto her memories of Ford, his great importance in the formulation of her artistic identity is clear. In this case, imitation was the sincerest flattery, but led to some harm as well as achievement.

The Tates' life in New York in 1926 and 1927 was not, however, all serious, high-minded dedication to Art. They also had a great deal of fun with their friends. At all the Tates' residences, Caroline was known for her ability to furnish a house and create delicious meals out of practically nothing. Andrew Lytle still remembers an omelette she concocted with okra and tomatoes at the Bank Street apartment.[100] Lytle, a student at the Yale Drama School, had been introduced to Allen through Ransom. He would join Tate and Ransom in their Agrarian Symposium and begin a career as a distinguished writer of fiction. In the 1930s, he would work closely with Caroline when they were both writing historical fiction. But in the 1920s the impressionable young Lytle once saw a visitor calling for Katherine Anne Porter at the Tates' Bank Street apartment. Allen answered the door. "The visitor was received with grave decorum and told, with a bow, 'The ladies of this house are at the riot in Union Square.' "[101] The poet Mark Van Doren recalled that at Bank Street, he and his wife "met [the Tates'] literary friends, who then, as always were legion; once we brewed with them a barrel of wine and called it sherry."[102]

The Tates lived even more closely with their legion of friends when they moved to 571 Hudson Street, affectionately known as Casa Caligari or Caligari Corner in tribute to the horror movie. Josephine Herbst, a denizen, said that "creaking up the stairs you half expected to see a skeleton wag from the ceiling."[103] To Malcolm Cowley, a frequent visitor, it was a "rabbit warren," and Caroline Gordon described it as a "fine pre-Revolutionary tenement" with "no hot water."[104] Dorothy Day, a journalist who would found the Catholic Worker movement, lived there with her small daughter. Her work and her way of life would have great impact on Caroline when both had converted to Catholicism two decades later. The short story writer Katherine Anne Porter, who would remain Caroline's friend for decades, lived in a room which was "a domestic pavilion with gingham curtains at a window [and] a flowering primrose," as Josephine Herbst described it.[105]

As this list of inhabitants indicates, although Allen still met with "the boys" outside of Caligari Corners, Caroline had found her own circle of friends within. Cowley remembered:

> Caroline, Katherine Anne Porter, and Josephine Herbst were the three talkingest women in New York. There was one night, one afternoon, that I certainly remember about. I had a date with Della [Dorothy Day's sister] for four o'clock, but she said phone me first, and I phoned her, and Della reported, "They're still sitting at the breakfast table. They just made another pot of coffee." Then I called in fifteen minutes more and they were still at the breakfast table, and I walked around, and I didn't call this time for half an hour, and Della said with relief, "They're just putting on their coats," so at last I could appear.[106]

As Caroline's later correspondence with Herbst indicates, she could talk to these women, all published writers, about the progress and problems of her fiction. Although their presence may have been a temptation to talk rather than write, their example as woman writers must have helped Caroline find her own identity as a writer and extricate it from her identity as Allen Tate's wife.

The Tates gained another charming and witty companion with the return of their daughter Nancy. Mrs. Gordon was suffering from breast cancer and her case was considered hopeless, so Nancy rejoined her parents. Nancy's presence increased Caroline's responsibilities, but the two year old's unconsciously prescient observations provided her parents' with great delight and a stream of comic anecdotes. Many of Allen's friends became Nancy's "courtesy uncles." As Caroline wrote Sally Wood, "Her courtesy uncles prize most the remark made to Donald Clark. She showed him a hideous headless doll and said 'This is my big baby, Donald.' 'Oh,' says he, 'Parthenogenic, I suppose?'—'Naw,' says Nancy, with equal aplomb, 'Santa Claus done it.' "[107] The Tates often told the story of how one day Nancy approached Allen at his typewriter and asked him what he was doing. "Making a living," he told her. Picking up a sheet of manuscript, she commented, "Mighty thin one."[108]

Despite Nancy's trenchant observation, the fortunes of the impecunious Tates were somewhat improved. As a sort of potboiler, Allen contracted to write a popular biography of Stonewall Jackson. In the summer of 1927, he and Caroline toured Virginia battlefields, sleeping in tourist camps, as Allen pursued his research and his increasing interest in Southern history. On its publication early in 1928, the biography certainly seemed to make the pot boil; according to Caroline, the book "sold three thousand in three weeks, enough for two years living," on a modest scale.[109] Allen's first collection of poems was also published in 1928. *Mr. Pope and Other Poems* may not have contributed much to the family larder, but it certainly advanced Allen's reputation as a poet. Finally, through the good offices of Ford Madox Ford, Allen received a Guggenheim Fellowship that would allow him to take his family to England and France and experience the expatriate society Hemingway had depicted in *The Sun Also Rises* and that he had heard about from his many friends who had already been abroad.

Caroline badly needed the break from New York for she was feeling overburdened by too many conflicting demands on her time. In the fall of 1928, she wrote, "I have come very near losing my mind this summer, what with family troubles and all."[110] She was troubled by her mother's illness and her mother's insistence that Caroline and Allen pursue their careers, including the trip abroad. Nancy Gordon wrote: "I shall probably be living when you get back and long after. I am ready to go and everyone else is ready for me to depart. We can write to each other better than we can talk." She then offers to pay for Caroline's dental work since her teeth "looked to me like they needed attention when you were here."[111] Although her mother's attitude had its generous elements, Caroline may have felt that her mother was saying "I don't need you or a last farewell from you," with the dental work another version of "doing my duty by you" when love is wanted.

Caroline also felt that if unwanted in one place, she was too much in demand in another, and her time was frittered away. "I have little to show for two—or is it three years—work. This book I have just finished [an unpublished mystery novel], and three fourths of another novel [probably *Penhally*], and two stories. That is all I have been able to accomplish." She blamed her inadequate output on the Tates' constant stream of visitors. "It is these young poets from the South—they call us up as soon as they hit Pennsylvania Station and they stay anywhere from a week to a month. I have gotten bitter about it."[112] With a larger cast of poets, this scenario reenacts the situation with Hart Crane in the Tory Valley; a visitor is eventually perceived as an invader. In this instance, however, the outcome was reversed; the Tates decamped, not the poets.

On September 28, 1928, Caroline, Allen, and three-year-old Nancy sailed for England on the United States line. As they went up the gangway, Allen, prepared to be a gentleman and a scholar, carried his grandfather's gold-headed cane and two volumes of *The Rise and Fall of the Confederacy*. Caroline's entrance was less auspicious, with Nancy by one hand and a big baby doll in the other. As if responding to these cues, a fellow passenger, one of a group of Rhodes Scholars, "had the effrontery to refer to me as 'the wife,' " Caroline wrote.[113]

To stretch their grant money, the Tates had booked for "tourist third," but a dearth of first-class passengers allowed them to use first-class cabins where they were "confortable and well fed."[114] A

violent storm ended this comfort, and, typically, Caroline turned it
into a joke at masculine expense. She wrote to Sally Wood: "Just
before we were landed I was requested to sign a letter of eulogy of
the captain who stayed on the bridge three days & nights and like
Allen, never changed his clothes. I was surprised to find that the
eulogy was the work of Allen—surely the most florid piece of
writing he's everdone."[115]

On arrival in London, the Tates set out to see the sights—
Westminster Abbey, the Temple, Magnus Martyr, Warwick Castle, and Stratford-on-Avon—until Allen was felled by the flu in
mid-October. The Tates' responses to England were determined
by their roles and expectations. For Allen, England meant intro-
ductions to London men of letters, such as F. S. Flint, Herbert
Read, and Harold Monro. Of course, he looked forward to meeting T. S. Eliot, with whom he had corresponded and to whose
poetry his own was often compared. He had anticipated his mirror
image, as he wrote to Donald Davidson: "I expected to see a small
man with an extremely intellectual face, yet lacking those features
that signify Will." Instead, he found that Eliot's "character, far
from being weak, is almost overdeveloped." Eliot remained mysterious to Allen, "sphinx"-like, but he found him "most amiable," a
judgment in which Caroline concurred. Although some of the men
were interesting, the women, he wrote, "are appalling. Great God!
What feet and ankles!"[116]

Caroline's expectations were different. She believed in advance
that she would not like the English, writing jokingly to her friend
Virginia Tunstall that "I, being a Gordon and nurtured in Celtic
prejudice . . . am all prepared to hate England. I've been hating
England and things English for thirty-two years. I shall not give up
that warm emotion for a mere stay of two months."[117] Caroline
also had a less whimsical reason for expecting to dislike England.
"I could afford a nurse for Nancy in France but not England. Six
weeks, with no interruption would enable me to finish this novel
I've been working on now for three years [Penhally]—but I see
little chance of getting six weeks. I am getting bitter about it."[118]

As it turned out, both Tates obtained a good place to work when
they moved to Oxford at the end of October. Caroline said that
she was able to rewrite half her novel.[119] Allen had decided on
Oxford for its beauty and the convenience of its libraries since he

was now at work on a second biography, that of Jefferson Davis. In addition, Oxford meant the companionship of Robert Penn Warren who was there at work on his biography of John Brown.

Joining the Tates' circle in Oxford was their friend, Leonie Adams, whose first volume of poetry, *Those Not Elect* had been published in 1925. She would also accompany them to Paris where she would share their accommodations until she returned to the United States in March. Although this situation may seem odd after Caroline's complaints about poets descending on her New York household, she was in some sense turning the tables on Allen in a way she would continue throughout their married life. Caroline needed her own circle of friends and dependents or she would always have her friends through Allen, as she did at the beginning of their marriage in Greenwich Village. She may also have been trying to recreate the extended family with its frequent long-term visitors that she had enjoyed during childhood summers at Merimont. Her comment after several months of Leonie Adams' company is revealing: "She is an awfully nice person, my favourite of all the lady poets, I believe. She doesn't have any of the proverbial vices of that craft."[120] Leonie was a fellow artist but did not have "proverbial vices" of male poets such as the Fugitives, Hart Crane, and other friends of Allen, as well as Allen himself.

Allen seems to have been puzzled by Caroline's tendency to extend their family group, sometimes to those he disliked. In the case of Leonie Adams, Allen could remain fond of their guest. As did Caroline, he respected her as a truly kind and good person; after several months of her company, he would write that her sterling qualities even withstood the test of time.[121] Her presence may also have been acceptable because he did not consider her a serious poetic rival. In a review of *Those Not Elect* two years earlier, he had written, "Miss Adams is personal, meticulous, detached from ulterior literary motives; a distinguished limited sensibility, she is a distinguished minor poet."[122]

On Thanksgiving Day, 1928, the Tates and Leonie Adams left Oxford for France in a chill fog. The next morning they made the Channel crossing from Newhaven to Dieppe, the longest but cheapest route. They then proceeded to Paris where Ford Madox Ford had made reservations for them on the second floor of the Hotel de Fleurus, a ten-minute walk from his apartment at 32 rue de

Vaugirard. At first, their plans were uncertain; they toyed with the idea of visiting Switzerland, but decided that the resorts and the trainfare would be too expensive. They were, of course, loathe to leave the Paris of the 1920s which would be celebrated by numerous expatriate writers. As Allen later wrote, "I was about to plunge into the French experience which young Americans in the twenties thought they must have or remain provincials."[123]

Ironically, their first contacts were familiar, not foreign. Robert Penn Warren joined them for two days on his way to Cannes, travelling, wrote Caroline, with a "South African youth named Chaumont Pierre de Villiers who did not know a word of French, in spite of his Huguenot blood."[124] They spent time in the cafes frequented by the expatriates, such as the Dome and the Rotonde. To Caroline they were "a sort of super-Greenwich Village. They actually appall."[125] She was also preoccupied by her more distant past, as represented by her fatally ill mother in Kentucky. Caroline wrote to her frequently, recounting Nancy's bright sayings and preparations for Christmas: "She is very much excited . . . but a little disturbed that it is Little Lord Jesus' birthday, and not hers; she is afraid he will get all the toys." She also told her mother of her visit to the Louvre which she found bewildering in its profusion. The one exception was the collection of Italian primitives, which she liked immensely and would later emulate in her own paintings.[126] Ford Madox Ford made the Tates and Leonie Adams regulars at his Saturday evening soirees, where, according to Allen, "we sat at the feet of the master; Ford, however, talked little but rather nodded agreement or disagreement."[127]

Ford did more for the Tates than allow them to sit at his feet. He was leaving for America in January and would give them his apartment near the Luxembourg Gardens rent-free in exchange for Caroline's retyping a five-hundred page manuscript for him. Caroline wrote, "He had it typed by a French typist and every e came out a and every inanimate object with sex, etc. I think, though, that I finally got it in shape."[128] In his memoirs, Allen described the "Spartan frugality" of 32 rue de Vaugirard: "[Ford's] flat consisted of a *petit salon* furnished with a divan which, like that of the typist in 'The Waste Land,' became at night a bed; this room, the British phrase, was the bed-sitting-room of my wife and me. At the far end was a small room with a small bed occupied by our daughter; near the entrance a narrow closet was just large enough for Miss Adams." The flat also contained a bathroom where the toilet was a hole level with floor, "requiring," remembered Allen, "considerable acrobatic agility."[129]

Despite these rigors, Caroline discovered that "Life certainly is much easier." The exchange rate was very favorable for Americans. Clothes were comparatively inexpensive, and Caroline set about finding outfits that could last her some years at home. Liquor was also cheap, and not subject to Prohibition as in America. Although Caroline had always worried about being too thin, now she wrote to Josephine Herbst, "I actually worry about my figure"; she had gained fifteen pounds. Although French food and wine must have contributed to Caroline's figure and state of well-being, an absence of household cares undoubtedly helped as well. She exulted, "We now have a femme de menage who does everything—even takes Nancy for a walk in the gardens—for twenty dollars a month."[130]

Allen and Caroline did not become as well-acquainted with the French as Nancy. Allen recalled, "I met in my first year in France very few French people, writers or mere citizens, and that was true of most Americans in Paris."[131] Caroline could never master the language; she was able to learn to read and write foreign languages, such as her Latin and Greek, but not to speak them, except for a few vital everyday phrases. In Paris she justified her failure to learn French, saying "It upsets me to even think about it when I'm trying to write."[132] To her chagrin, she found that the language had its revenge. She remembered "the look on a shopkeeper's face when, in the course of buying a crib for my daughter, I asked if a *matelot* [sailor] went with the bedstead, meaning of course a *metelas* [mattress]."[133] Young Nancy managed best of all and became fluent, perhaps too fluent for these Southerners. "Her favorite book is still 'Battles and Leaders of the Civil War,' " wrote Caroline to Sally Wood, "but now the other day when she was looking it over she observed 'There's Monsieur Stonewall Jackson.' We can't have that."[134]

Portrait artist Stella Bowen depicted the Tates as they appeared in Paris in 1929. She lived in another apartment at 32 rue de Vaugirard with her and Ford's daughter Julie, who was several years older than Nancy Tate. Caroline's description of the portrait seems prophetic: "Allen and I, held together in space, by Nancy, as it were."[135] Indeed the Tates' marriage would begin to deteriorate seriously when Nancy left them to marry and she no longer served as a buffer and common goal.

In February 1929, the Tates were together and beginning to enjoy Parisian life when their happiness was marred by bad bouts with the flu. First Nancy ran a high temperature, then Caroline began to suffer in her nose and throat, and, finally, Leonie sickened and developed an abscess in her ear. Allen, as the only non-invalid, was in charge of caring for the others, a particularly worrisome task considering that the toilet was parallel to the floor so he expressed concern about the women's balance.[136] Further comic relief came from Hart Crane. Caroline wrote to Josephine Herbst, "He seems to be enjoying Paris hugely, and cheered our semi-invalidism with tales of his going on at the Coupole—how Kiki (I cannot call her Mrs. Man Ray) slipped her left breast out of her decolletage and wagged it at him by way of greeting."[137] Although Allen was spared this round of illnesses, he would have a number of severe colds throughout their sojourn in France, as would the others.

Perhaps because of this onset of illness, Allen's thoughts turned toward religion. He wrote to Donald Davidson on February 18, "I am more and more heading toward Catholicism. We have reached a condition of the spirit where no more compromise is possible. That is the lesson taught us by the Victorians who failed to unite naturalism and the religious spirit; we've got to do away with one or the other; and I can never capitulate to naturalism."[138] Caroline's thoughts were necessarily more mundane. Reflecting on France after her return to New York, she wrote: "I must say let all of us that can turn Catholic at once. Yes, you must have plenty of servants to abandon yourself to your emotions, even to what mind you have. I can't work when I'm doing scullery work because even when I get the time my mind won't take hold of any problem."[139] Although the Tates would not convert to Catholicism for another twenty years, their approaches to it remained constant: Allen was looking for an answer to an intellectual dilemma, while Caroline was seeking a way of life.

Despite the distractions of a new country, their friends, and illness, Allen and Caroline did attempt to write. Allen seems to have accomplished more. Caroline wrote on February 11 that "he's been writing a good deal of poetry; some of it entirely unlike anything he's done."[140] He also spent several mornings a week at the American Library, working on his biography of Jefferson Davis.[141] When he confronted a publisher's deadline for the book, Caroline remembered, "I had to drop my work for a whole month to help Allen," mainly with typing. The book was finished in July and the Tates celebrated by "getting drunk" with Hart Crane.[142]

Caroline was accomplishing less for the obvious reasons: familial responsibilities, although Nancy did have a nurse; her mother's death on January 22; a new way of life; and inconvenient visits from the Tates' ever-increasing circle of friends. She recognized, however, that the reason for her inability to work enough was more profound, though common to many writers. That March she wrote, "I believe that aversion to work is so deep-seated in most of us that we will clutch at anything that distracts, and the only practical plan is to get everything like that out of the way before you start in, just as you sweep the room before you sit down at your typewriter."[143] Over the next decade, she would confront this problem with increasing professional dedication and achievement, but she could never completely conquer it. Her feeling that writing was, as she put it, "torment," meshed too neatly with the demands placed on her as a woman: wife, mother, and hostess.[144]

She was able to work more effectively, however, when the Tates left Paris in mid-July after handing the keys of 32 rue de Vaugirard back to Ford. Still accompanied by Leonie Adams, they went to Concarneau on the coast of Brittany to spend the summer. The Tates stayed at Le Grand Hotel de Cornouailles and Leonie Adams obtained accommodations at a nearby *pension.* Allen's somewhat bleak description of the summer may reflect his sadness over the death of his mother on July 17: "a chilly summer on a rough beach where the sea was cold, the food of the Grand Hotel greasy, and only the blue, red, and green sails of the sardine boats lent color."[145] This lack of distracting charm seems to have helped Caroline since she produced several short stories. That fall she wrote to Josephine Herbst that "the devil somehow got into me while I was in Brittany—but I swear it will be the last I ever write. I simply loathe short stories. They are all just a trick, and they simply drive me mad."[146] Although her short stories won great acclaim and are often considered superior to her novels, Caroline Gordon always regarded herself as a novelist and the short story as a lesser genre.

The Tates and Leonie Adams returned to Paris in the fall. Allen thought they should move to London where he could have access to the British Museum for his third biography, the never completed, *Robert E. Lee,* but, as Caroline said, "we simply couldn't leave Paris."[147] The Tates took rooms at the Hotel de la Place de l'Odeon and Leonie Adams at the nearby Corneille. Madame Gau, an old Frenchwoman, was placed in complete charge of Nancy. Caroline was further relieved by frequent invitations to dinner from Ford, an excellent cook, often followed by evenings at the Deux Magots or Closerie des Lilas, where Ford imbibed the several cognacs he felt necessary for a night's sleep. Despite attacks of flu in November, the fall of 1929 seems to have been a happy one for the Tates.[148]

Interesting friends, old and new, lent much to their enjoyment. Allen's old friend from New York, the poet John Peale Bishop, had married a wealthy woman and was living in some state at the Chateau de Tressancourt at Orgeval, twenty miles from Paris. When the monied Bishops met the impecunious Tastes for dinner, "compromise" restaurants were sought. John Peale Bishop was also rich in friends, despite his somewhat diffident manner. With Robert Penn Warren the preceding fall, Allen had urged Bishop to publish his stories, *Many Thousands Gone* and his poems, *Now With His Love.* Fortunately for the Tates Bishop's most notorious friend was his old Princeton classmate, F. Scott Fitzgerald, currently in Paris with his wife Zelda and daughter "Scotty."

The Tates met the Fitzgeralds at a party given by the Bishops. Allen remembered Scott making frequent trips to the kitchen where he had hidden a bottle of gin. Perhaps these forays account for the nature of their first conversation. According to Allen, shortly after Bishop introduced them, Fitzgerald asked him, "Do you enjoy sleeping with your wife?" The incredulous Allen asked him to repeat what he said; when Fitzgerald obliged, Allen replied, "It's none of your damn business."[149] The attention-seeking chronicler of the Jazz Age and the courteous Southern gentleman later became better friends, in spite of this initial encounter, but their contrasting personalities would never allow them to become particularly close. According to one of Fitzgerald's biographers, Andre Le Vot, Fitzgerald's conversations with staunch Southerners Tate and Gordon "confirmed Scott's antipathy toward the industrial and mercantile society he had already criticized in his interviews" and led Scott to choose Maryland as his next home in the United States.[150]

When asked about Scott Fitzgerald late in life, Caroline said little about him as a person, but praised his art as a novelist, particularly in *The Great Gatsby*. She had more observations about the famous novelist's wife whom she characterized as a "Southern belle," who lacked enough pride to prevent her from dancing with gigolos at dance halls. Caroline also spoke of Zelda's fierce competitiveness with Scott as she strove to establish her own career as a writer, dancer, and painter. Caroline did not believe that Zelda's ambitions helped destroy Scott, but that she "just wanted to have her own accomplishments."[151] In the Fitzgeralds she saw a more glamorous and more tragic version of the tensions between two working artists in her own household.

The Tates met Fitzgerald's sometime friend and rival, Ernest Hemingway, through the offices of expatriate American Sylvia Beach, publisher of Joyce's *Ulysses* and proprietor of the now legendary Paris bookstore, Shakespeare and Company. After their introduction, Allen took a walk with Hemingway. As in his first meeting with Fitzgerald, questions of virility came to the fore, in a somewhat displaced way. According to Allen, Hemingway asked him if he knew that Ford was impotent, to which Allen sardonically replied "that that must be very sad for Ford, but not being a woman I could feel very little interest in Ford's sexual problems."[152] The basis for Hemingway's remark, though typical of his outwardly macho personality, was laid much earlier in one of Allen's reviews of his work. On February 14, 1927, Hemingway had written Maxwell Perkins, his editor at Scribner's, that "Up in Michigan," a short story, "is publishable and might set Mr. [Allen] Tate's mind at rest as to me always avoiding any direct relation between men and women because of being afraid to face it or not knowing about it."[153]

Tate and Hemingway did become close friends in that fall of 1929. They attended bicycle races together on Sunday and spent evenings in cafes, sometimes also with the Fitzgeralds and Ford. Hemingway, the sportsman, did not seem to mind Allen's lack of

interest in physical prowess, but praised him for his "moral courage" when Tate would not go skiing with him.[154] Since Allen would
not compete on Hemingway's turfs, sports and fiction, Hemingway
was not threatened by him and could respect and like him as
"damned intelligent and a very good fellow."[155] For his part, Allen
regarded Hemingway as a Southerner manqué. To his fellow
Agrarian, Donald Davidson, he wrote: "Hemingway, in fact, has
that sense of a stable world, of a total sufficiency of character,
which we miss in modern life. He is one of the most irreconcilable
reactionaries I have ever met; he hates everything that we hate,
although of course he has no historic scene to fall back on."[156]

Caroline Gordon did find common ground with Hemingway in
sport, since her beloved father shared that enthusiasm. She wrote
to Josephine Herbst that Hemingway "is trying to persuade us to
settle in Arkansas—Pauline's [his second wife's] people have a lot
of wild land down there. It sounds fine for hunting and fishing."
She remembered living as a girl in nearby Poplar Bluff and her
father "spending all his time on the river."[157] Despite these affini-
ties, Caroline was not always present at these encounters with
Hemingway or other literati. For instance, Hemingway wrote to
Perkins on 15 December 1929, "got shaved and to Mass with Pauline, Allen Tate, and a couple of citizens."[158] Caroline was not
sharing Allen's flirtation with Catholicism and also needed time to
herself for her writing and household tasks, or perhaps Hemingway did not consider her important enough to mention in the
letter.

If Caroline sometimes kept herself apart from the circle of "the
boys," her peripheral status as a woman was confirmed at Gertrude Stein's weekly gatherings. The Tates had met Stein the previous fall; on the day of their arrival in Paris, Allen received a
characteristically imperious note from her: "You and your wife
will come to tea on Thursday at 27 rue de Fleurus."[159] In her room
filled with the works of Picasso and other modern artists, Allen
remembered that "the ladies were, of course, second-class citizens,
and segregated around a table to be entertained by Miss Toklas
[Stein's companion], while Miss Stein engaged the men"[160]

Perhaps because she preferred laughing to crying, Caroline remembered these Thursday *salons* as comic anecdotes, particularly
one at which Ford, the Hemingways, and the Fitzgeralds were

present, and Stein was "explaining that she was 'the flower of American literature: first there was Emerson, next there was Whitman—and now there is *ME*.' "[161] Men might have some status, but Stein outranked them all, including Allen. She informed him that, "being a Southerner, he could not be expected to know any history!"[162]

On Saturday evenings, the Tates faithfully attended another *soiree* at Ford Madox Ford's apartment. Also frequently present were the Canadian writer Morley Callaghan and a poet from California, Howard Baker.[163] Sometimes, Caroline remembered, they would play a game that involved assigning their acquaintances to the appropriate circle of Dante's Inferno.[164] Allen recalled another game, *bouts-rimes:* "Ford passed around pencils and paper and assigned us the rhyme words. . . . Everybody had to use the assigned rhymes, and the winning sonnet was graded on both its quality and the speed of its composition. Ford himself usually won and gave as a second prize a melancholy-looking round cake which was sliced at once and eaten with moderate zest by the company." Allen added although they were "bored," they participated "because we respected and loved the Master, who was disinterestedly kind to those who disliked him, as well as to those who loved him."[165]

In the case of Caroline, Ford's love may have been a bit more ardent than that of a disinterested mentor. Despite his ungainly figure and rather homely face, Ford fancied himself quite the ladies' man, and his gallantries extended to Caroline. Although she was married to Allen and he to Stella Bowen, he proposed to Caroline in a Parisian church, of all places. She recalled that she told him he was "crazy." To this interviewer, she added, with a gleam in her eye, "he should have been ashamed of himself." Sally Wood, who was with the Tates and Ford on various occasions over the next few years, remembers that this affectionate refusal to take Ford seriously was Gordon's characteristic response.[166]

A letter he wrote her on the Tates' return to America, the only one Caroline kept in her papers, gives some indication of his approach. He tells her that when he reaches America, he will go to her immediately, even if she refuses him, and waxes sentimental over how dear her birthplace would be to him. The prospect of his next sight of her, he adds, is what gives his life meaning.[167] Ford, the novelist of his own life, may have enjoyed playing the devoted courtier to Caroline's chaste lady.

In addition to his gallantries, Ford did Caroline the much greater service of helping to transform this busy, easily distracted, insecure, unpublished novelist into a professional writer. In her last months in Paris, in the fall of 1929, Caroline was increasingly frustrated at her inability to finish a novel and get it published. She further distracted herself by planning other projects, including a biography of her ancestor, the explorer Meriwether Lewis, in the hope that it would be "a pleasant way to make a little money." She knew, however, it would be just another delaying action. To Josephine Herbst, she wrote, "Almost everybody has a book out this fall. . . . I am getting embarrassed about my output."[168] In desperation, she finally showed her fragmentary manuscript of *Penhally* to the kindly Ford, who later wrote that it gave him a "shock of delight at the beauty of the writing and the handling of the material."[169]

More valuable than Ford's praise, however was his discipline, as Caroline wrote to Sally Wood.

> Ford took me by the scruff of the neck about three weeks before I left, set me down in his apartment every morning at eleven o'clock and forced me to dictate at least five thousand words, not all in one morning, to him. If I complained that it was hard to work with everything so hurried and Christmas presents to buy he observed "You have no passion for your art. It is unfortunate" in such a sinister way that I would reel forth sentences in a sort of panic. Never did I see such a passion for the novel as that man has.[170]

Her father, Professor Gay at Bethany College, Allen Tate: Caroline Gordon's life contained a succession of male mentors who at once enabled her and yet crippled her by her dependence on their opinion. In Ford Madox Ford, Gordon had found another male mentor, but one with the gift Caroline needed most at that time, a "passion for the novel" that led to professional dedication and discipline. On the Tates' return to America, Caroline would establish herself as a published writer and enter her most productive decade.

CHAPTER 4

When they arrived in New York in January 1930, the Tates began to reestablish their pre-Paris life of temporary lodgings and hand-to-mouth freelancing. They found a furnished apartment on Twenty-Sixth Street, near Ninth Avenue, which was so small that Caroline declared, "I debate about buying a tea strainer because I don't know where I'd put it."[1] She did make one significant change from their former New York lifestyle when she hired a daily cleaning woman. Caroline now believed she was entitled to put her writing first; she had returned to New York a published author. In November her short story "Summer Dust" had appeared in the first issue of *Gyroscope,* a mimeographed little magazine published by the California poet and critic Yvor Winters.

Appropriately for her first publication, "Summer Dust" is a portrait of the artist as a young girl, here called Sally, a name Caroline would often use for her fictional self. In this story, Caroline returned to the Merimont of her childhood to explore a Southern girl's increasing recognition of sexual and racial prejudices. "I'm not a nigger," Sally realizes, as she looks at her black playmate Son.[2] She escapes society's constrictions by retreating to the fantasy world of *Green Fairy Book* or by withdrawing to the woods where she can create her own fantasies. Significantly, she imagines a gypsy woman, "a dark woman with a crimson scarf bound tightly about her head, and two round gold earrings as big as saucers, swinging from the scarf as she walked." In her nonconformity and untrammeled self-expression, the gypsy woman suggests Sally's vision of her future self, woman as artist.

Caroline, however, does not end "Summer Dust" on this positive note. Sally runs away from an old black woman after her
brothers tell her that the woman eats blood. With one stroke, they
have managed to reinforce racial prejudices, break any bonds of
female solidarity, and assert their control over her imagination.
The lesson of a woman's place is further reinforced when her
brother tells her that "She who calleth her brother a fool is in
danger of hell fire." Religion supports the male, and as Sally next
discovers, so does law. Her cousin has just returned from court
where he boasts of his success in evading the acknowledgment of a
poor white girl's child as his own. Sally escapes unbearable reality
by returning to her fantasies from the *Green Fairy Book*. In "Summer Dust," Caroline Gordon is suggesting that the artist's imagination can be her refuge and her strength as in Sally's image of the
gypsy woman, but that the power of the patriarchy can taint her
imagination, as in her brothers' story of the black woman drinking
blood. Her own imagination becomes suspect and frightening to
her, hence Sally's passive retreat to the *Green Fairy Book,* a fantasy world created by others.

Caroline returns to questions of racial and sexual roles in her
second published story, "The Long Day," which appeared in February in the second issue of *Gyroscope*. In this story, she explores
these issues from the masculine point of view. A small white boy,
Henry, wants to go fishing with his black mentor, Joe, but his
mother is hesitant about granting permission. Henry's Uncle Fergus laughingly reassures her that Joe's wife, Sarah, "won't be up to
any more didoes today" since "Joe gave her a good larruping before he came up to the house."[3] Henry's mother replies, "I hope he
did. . . . I hope he beat her within an inch of her life." Uncle
Fergus refuses to take the situation seriously and comments, "Joe
likes his mammas hot. . . . Georgy was no sucking dove." Henry's
mother then asserts her world's standards of conduct. "Georgy
behaved herself very well while she was on this place. At any rate
she never attacked Joe with a razor. This razor business is too
much." Sarah is the victim of the sexual and racial double standards expressed by Henry's mother; Sarah's husband can have an
affair, but she can't retaliate with a razor; wife-beating is an acceptable practice for blacks with their lower standards.

The adults' conversation is incomprehensible to Henry who proceeds to Joe's cabin after repeated warnings from his mother about not entering the cabin. Joe keeps Henry busy outside the cabin all through the long day. At the end of the day, when the door is opened, Henry sees Sarah's bloody and self-mutilated body on the cabin floor. The story ends with Henry running "as fast as he could toward the house." Henry is running toward the safe hypocrisies and condescensions of his mother and uncle. Caroline has gruesomely illustrated the education of the white Southern male into racial and sexual prejudices similar to those of Sally's brothers and cousins in "Summer Dust." "Sally," of course, is the diminutive of Sarah, and Gordon may be suggesting that a grown woman's only power over herself is the ability to complete her education in self-annihilation.

Caroline's debut was readily recognized as auspicious. "Summer Dust" was selected for Edmund J. O'Brien's collection of *The Best Short Stories of 1930.* "The Long Day" was reprinted in *Scribner's* August issue, and Maxwell Perkins, legendary editor of Fitzgerald and Hemingway, asked to see more of her stories.[4]

Not all was triumph, however. Another short story, "Funeral in Town," was rejected by Bernard Bandler of *Hound & Horn* because he considered the main character vague and many of the details irrelevant.[5] "Funeral in Town" would have a further history of rejections and remain unpublished. As she apparently wrote to Ford Madox Ford, Caroline was also worried because her novel was nowhere near completion. In his reply, Ford insisted that she finish the novel and attempted to allay another of her fears, that her work was too derivative of his. He pointed out that her strongly individualistic character stamped her work and that his influence was found merely in matters of technique.[6]

Caroline, with Allen, was also confronting the perpetual problem of how to make a living while continuing to write. Never a lover of urban life, at the beginning of the Depression she found New York "terrible" and longed for a home in the country.[7] As in the case of the Tates' earlier move from the city to Patterson, they believed they could stretch their earnings much further in the country. This time they wanted the South to be the site of their country home. Over the past few years, Allen had become increasingly obsessed by his Southern heritage, as demonstrated in his biographies of Stonewall Jackson and Jefferson Davis, as well as in his poetry, most notably the "Ode to the Confederate Dead." Caroline's imagination was also engaged by the South of her childhood in her short stories and of her family's past in *Penhally*. A further impetus was supplied by Caroline's widowed father who was tired of his ministries in the towns around Cadiz, Kentucky, and longed for complete freedom to pursue sport. He hoped to contribute to a down payment and use the Tates' home as a base for his fishing expeditions.[8]

After some discussion of Virginia as a possibility, the Tates decided to live at Caroline's grandmother's farm, Merimont, while they looked for a home in Kentucky or Tennessee. They left New York at the beginning of March and headed South.

The Tates considered building a cabin on the Merimont land, but they fell in love with an antebellum home on a bluff above the Cumberland River with a spectacular view of Clarksville across the river. Caroline described the hill to Sally Wood: "It is shaped just like a crouching lion. The lion's head and shoulders front the river and the house sits on, say, his forehead. One drives up his spinal column, right to the front door; the ascent is gradual and you do not realize till you get on the big porch how high up you are. It is swell, at night, when the lights in the town come out."[9] The house was constructed of brick, once painted white, but with a rosy hue where the paint had worn away. It had three floors, linked by a steep and narrow staircase. The bottom floor was cut into the hillside so that it was a ground floor on the river side and a basement floor on the other side. This lowest floor contained a kitchen, small bedroom, and a dining room that occupied one complete side from front to back. The upper floors had parlors and bedrooms, opening out on to the porch. Cedars, crepe myrtles, and altheas grew in profusion on the grounds.

The house would need much restoration and modernization, including electricity. Caroline's father decided he wanted his freedom to search for the best fishing areas, which did not include the Cumberland, so he dropped out of the scheme. Just when it appeared that the house was beyond the Tates' very slender means, Allen's older brother, Ben, the successful businessman, offered to buy and restore the place for them. Ben believed that Allen needed a permanent home for his writing and that the Tates could then support themselves once such a retreat was provided. In honor of his generosity, and optimism, the place was called Benfolly.

Since Benfolly came with one hundred acres and could be a working farm, its purchase is often considered a result of Allen Tate's participation in the Agrarian movement. In 1930 some of the former Fugitives, Tate, Andrew Lytle, John Crowe Ransom, Robert Penn Warren, and eight others, including non-Fugitives, published a manifesto, *I'll Take My Stand*. They identified themselves as latter-day Rebels by borrowing their title from the Confederate hymn, "Dixie." The Agrarians believed they were taking their stand against a new and even more invidious Yankee invasion, the encroachment of Northern industrial society on the agricultural South. As the volume's Introduction states, the articles "all tend to support a Southern way of life against what may be called the American or prevailing way; and all as much agree that the best terms in which to represent the distinction are contained in the phrase, Agrarian versus Industrial."[10]

The distinction is more than sociological; for the Agrarians, agriculture had moral and intellectual implications: "an agrarian society is one in which agriculture is the leading vocation, whether for wealth, for pleasure, or for prestige—a form of labor that is pursued with intelligence and leisure, and that becomes the model to which other forms approach as well as they may. The theory of agrarianism is that the culture of the soil is the best and most sensitive of vocations" (xlvii). Allen Tate shared these views and placed them in the context of modernism. In a letter to John Wheelwright, he wrote that modern society kept the cultured classes removed from the basis of their livelihood (presumably the soil), and so formed an arid, attenuated type of mind that will not produce a great literature.[11] Tate, of course, is describing the negative aspects of his own "mind" as revealed in his early, modernistic poetry, and he may have hoped that a life closer to the source would enrich his poetry as well as his life.

The Agrarians knew that the days of the old plantation were over; they realized that the planter's life of cultivated leisure had been supported by the lack of leisure of his cultivating slaves. In 1931, Tate wrote that the Agrarians wanted the continuation of a tradition which he believed had been destroyed by the North; they did not wish to return to the days of plantation slavery.[12] The

Agrarians believed that the self-sufficient yeoman farmer, not the large-scale planter, could regain this "continuity of tradition" and reap the moral and intellectual benefits of an agricultural way of life. This ideal of the self-sufficient yeoman farmer was not original to Tate and the Agrarians, but derives from America's origins, when it was most notably expostulated by Thomas Jefferson with his fervent belief in the tutelary benefits of nature and the pernicious influence of cities.

I'll Take My Stand was a manifesto, and like most such documents was more honored in the breach than in the observance. As John Crowe Ransom later said of the Agrarians, "Perhaps there was only this hollowness, that like gentlemanly conspirators in a movie we tended very kindly to conceal from each other the fact that individually we had no expectation of throwing up a dyke as would turn a historic tide from overflowing our region as it had submerged the others in the land. We were engaged in a war that was already lost."[13] The authors of *I'll Take My Stand* were not the most likely soldiers in the Agrarian battle. They were intellectuals who supported themselves by their pens, not their plows, and, increasingly, in sowing knowledge in the minds of college students, not corn in the soil.

Accordingly, Benfolly was mainly farmed by a succession of tenants who produced corn and soybeans, while the Tates produced manuscripts. Allen was no more fond of gardening than he had been when the Tates lived in New York's Tory Valley. Caroline maintained a large kitchen garden with the help of hired hands. With such assistance, they had, at various times, chickens, ponies, a cow named Daisy Miller (more as an allusion to Edmund Wilson's novel than Henry James's novella), a calf called Uncle Andrew in Lytle's honor, and a series of cats and dogs.

Caroline's attitude toward Agrarianism was at once more frivolous and more serious than those of the brethren. She could and did mock them both collectively and individually, as in this boys-will-be-boys letter to Sally Wood: "The Symposiers [of *I'll Take My Stand*] are having a great time. An Ivanhoe suddenly appeared in the list, Stringfellow Barr, editor of the Virginia Quarterly Review, who writes an article in said review to the effect that industrialism is a swell thing, let's whoop it up in the south etc. This gives the

Nashville brethren a chance to accuse him of selling out. They then proceed to recriminate each other in the pages of the [Nashville] Tennessean, New York Times, etc. even brawling a bit on the A.P. wires."[14] Her husband did not escape her satire either: "Allen is developing the true landlord spirit. He told me that when they went to the second hand furniture store, Jesse [their tenant] kept demanding all kinds of fancy things, 'Like what?' I asked, thinking perhaps a radio. 'Like chairs,' says Allen sternly. 'I told him there was nothing doing. Why, Jesse, I said, you can sit on a box.' "[15]

Caroline's humor here is actually of the gallows variety. She believed "the cause" was more than "lost"; it was dead as a mackerel and had been for so long that its remains only served to taint the present, already unbearable in its vulgarity. She wrote to Lincoln Kirstein, an editor of *Hound & Horn,* "The focus of my feelings, of course, is regret for the lost cause. It would have been better, I think, if our grandfathers had been carried off the field dead. The South as it exists today has little of the Old South in it— we have sold out certainly."[16] Her vision is uncompromisingly bleak; no wonder she seemed to regard the Agrarians as boys playing at history.

Caroline's practicality also caused her to see the essential absurdity of the Tates' position as owners of a landed, columned, antebellum dwelling. In June 1930 she wrote to Josephine Herbst, her fellow denizen of the Hudson Street tenement, as she awaited an enormous shipment of Allen's dead mother's accumulated furniture and objects: "I am rather appalled at the prospect of taking care of them, although God knows what we would have done without them. It is really funny. Here we are, absolutely broke, even broker than usual, in a magnificent house—at least it seems truly magnificent to my eyes used to such sights as Caligari etc . . . with two baths! Of course there is no water for the baths. There wouldn't be."[17] The house's water was supplied by a cistern that would frequently be exhausted by droughts or a large influx of guests. Its behavior was an inevitable theme in the Tates' correspondence and seems a fitting symbol of two freelance writers trying to lead the life of plantation gentry.

Were it not for Caroline's relations, the extensive Meriwether connection, the intellectual Tates would probably have had little contact with the local community. Danforth Ross, one of Caroline's distant cousins, recalled that "few people knew much about the Tates in those days because they seemed to keep to themselves. If so, the feeling was mutual. The Tates went their way and the people of Clarksville went theirs." The local people found the Agrarians amusing, says Ross, "because they talked about the importance of farming and yet didn't know much about farming themselves." Allen Tate, on the other side, did not seem to think much of the community's knowledge of his turf. Dan Ross recollects that "My mother did seek to open a discussion with him about his poetry at a party she had invited him and Caroline to. He responded that he had written his poem for specialized readers and that he liked to talk about other things at parties. Naturally my mother was indignant and never forgave him."[18]

Because of her Meriwether kin, Caroline found it much easier to mingle and participate, albeit in the role of the family's rebellious young modern who had lived in Greenwich Village and Paris. A cousin, Norma H. Struss recalls:

> There were only two places to defeat the hot summer—Spring Creek, a lazy creek running in the middle of our neighborhood, and Dunbar's Cave, a few miles outside of Clarksville. . . . It was customary for all of us to gather at the creek almost every afternoon or the cave where nearly every summer family reunions were held. It seemed to me Caroline always rather dominated the discussions asserting her ideas and going into extensive arguments with opposing ones. Most of the family were still loath to accept 20th century ideas and Carrie was often impatient. Then she would shrug her shoulders and laugh and make some dry humorous comment. Allen came to the creek sometimes. He seemed much more reserved and preferred to speak quietly with individuals.[19]

By July, after five months of these discussion partners, Caroline was longing for a visit from Sally Wood: "It will be a godsend to have somebody to talk to besides the kin. I am getting pretty tired of them."[20]

The Agrarian brethren, many of whom were living in Tennessee, were one source of non-Meriwether companionship. The Tates made frequent trips to Nashville where they played poker at the home of Lyle Lanier and saw other friends and acquaintances. The Agrarians came to Benfolly too, both singly and collectively. Robert Penn Warren and his first wife, Cinina, were living in

Memphis and visited the Tates at Benfolly for Thanksgiving and gatherings of the brethren. Stark Young, New York drama critic and novelist, stopped by on the way to visiting his family in Texas in the summer of 1931, to Caroline's great delight. No mean conversationalist herself, she found him, "a very entertaining person, the most fluent talker I ever heard."[21] Among this array of the talented and witty, the Agrarian whom Caroline most enjoyed was Andrew Lytle, both for his kind nature and their common interests. After his studies at the Yale Drama school, he returned to his father's farm in Murfreesboro, Tennessee. With his car loaded with farm products, including turkeys, he would arrive for long visits at Benfolly. Despite these harbingers of pleasure, his visits were working ones since he was completing his *Nathan Bedford Forrest and His Crittter Company* (1931). Allen was on the verge of beginning his never-to-be-finished biography of Robert E. Lee, and Caroline was writing the Civil War section of *Penhally*. Indeed, the inhabitants of Benfolly could spend so much time in the past that a time machine would be superfluous. Sally Wood remembers a visit in which Lytle was "pacing about the house with blank eyes, giving military orders. 'Then General Forrest said,'. . . . Only occasionally did he become himself. Most of the time he actually was General Forrest."[22]

Although the Tates had left New York, it followed them in the persons of some of their close friends. In October 1930, Malcolm Cowley and his first wife Peggy stopped at Benfolly on their way to Mexico, and Cowley remembered this detail of Caroline at work. The pantry contained "case after case" of Coca-Cola, in the "stronger southern version" filled with caffeine, which Caroline would drink "to stay alive while working."[23] Although she enjoyed visitors, the stimulant was undoubtedly welcome as she tried to write, run the household, and entertain the company.

After the Cowleys came another group from New York, which included Ford Madox Ford, Susan Jenkins Brown, and Harold Loeb, the prototype of Robert Cohn in Hemingway's *The Sun Also Rises*. The party, recalled Caroline, spent "a grand time, going in swimming . . . drinking corn liquor, and talking." The group migrated to Nashville to "support" Ford as he lectured to a woman's club and attended a dinner at Andrew Jackson's home, the Hermitage; Caroline termed both events "terrible."[24] On their return to Benfolly, Red and Cinina Warren joined the houseparty with some other guests from Allen's Vanderbilt days.

Not all visitors were sympathetic to the Agrarians, their cause, and their way of life. In February 1931, Edmund Wilson, novelist and critic, arrived at Benfolly to evaluate the Southern way of life for himself and the *New Republic*. With him were Louise Bogan, the poet and poetry reviewer, and her husband, the poet and novelist Raymond Holden. Unfortunately, Southern hospitality initially failed, for when the group arrived unexpectedly, the Tates were visiting cousins at Summertrees, a nearby Ferguson farm, and the Tates' cook, Beatrice, was nowhere to be found. When the Tates returned to Benfolly, Caroline later wrote, "I found [Beatrice] locked in the cabin, dead drunk! She had a bad cold and when I left I told her to take a hot toddy. She must have taken a good one."[25]

The visit continued to be awkward and uncomfortable. Caroline never really liked Louise Bogan. To make matters worse, she had recently learned that Louise had told Leonie Adams "everything we ever said about her just before they started down and Leonie had written wanting to know if it was all true." Caroline's tongue could be equally mischievous, and she wrote of Louise to Sally Wood, "She is utterly irresponsible and a born alcoholic."[26] Caroline further described the exigencies of the visit in a letter to Josephine Herbst: "They wanted to go out to Andrew Jackson's home, a place I'd sworn never to go, so we went on a dripping day, and sat in the car passing the bottle round among signs which said, 'No picnic lunches' until we got quite tight." Then, appropriately for Caroline's feelings, the party spent a night at a Nashville hotel where they played the parlor game "Murder."

Wilson, too, did not entirely enjoy the visit. His journal records impressions of pretension amid decay as he made the rounds of the Meriwether connection's farms with the Tates. Caroline did not help matters any when she gave him a bit of a tobacco leaf to taste. She wrote gleefully, "It was mean, I know. . . . He went into a slight convulsion."[27] Wilson also felt that the Agrarian brethren were failing in hospitality toward him by using him as a scapegoat Yankee. He later wrote to Tate, "Did you or did not you and John Ransom and [Donald] Davidson have the abominable manners to sit around and entertain Raymond and Louise and me with a prolonged headshaking and jeering over an unfortunate who had presumed to come South and try to edit a paper, and with a sour account of other Northerners who had had the effrontery to try to hunt foxes in Tennessee [instead of Virginia where the sport was traditional]?"[28]

Wilson got his revenge in "The Tennessee Agrarians" which appeared in the *New Republic* on his return.[29] Caroline described the article and the Agrarian response in a letter to Sally Wood: "The symposium boys are all mad as hops, of course. He did a very nice impressionistic study of Cousin John [Ferguson] and Summertrees, then threw the agrarian symposium in along with the lavender and old lace, giving the impression that they were all sitting around the trees, reading Greek and brooding on their ancestors." Since she never believed the Agrarian cause was more than an intellectual diversion, Caroline took Wilson's article more philosophically: "It's simply that people can't see anything that isn't already in their heads. We took him out to see Mister Rob [her uncle Robert Emmet Meriwether] who, though he isn't as easily classified, is much more interesting than Cousin John, simply as a case. If he had been a starving miner, Edmund would have understood his case, but as he wasn't, he didn't see anything at all—but Summertrees is more the kind of thing he expects to find in the south, so of course he leaped on it."[30]

The Tates' first year at Benfolly included many other visitors such as Allen's father and brothers, the poets Phelps Putnam and Howard Baker, and many others. The Agrarian ideal of cultivating the intellect in rural peace did not work for Caroline. Although she may have avoided New York's continuous distractions by moving to Tennessee, she also acquired the burdens of an innkeeper. In the quiet intervals between guests, Caroline did manage to write and had the pleasure of working in a house and environs in which she took great pleasure.

In two stories written in the fall of 1930, Caroline presented her own analysis of the postbellum South, perhaps her fictive counterpart to *I'll Take My Stand*. In her first two stories, "Summer Dust" and "The Long Day," she had exposed racial tensions; in "Mr. Powers" and "The Ice House" she would examine the tensions between classes of whites. Jack and Ellen Cromlie and their daughter Lucy are the new owners of an estate much like Benfolly. Jack contracts with Mr. Powers and his family as tenants, and they move onto the place. As Ellen literally and figuratively looks down on their cabin from her hilltop, she frequently speculates on the Powers family. She, however, is only comfortable with them on an employer-employee basis: she considers offering Mrs. Powers some of her spare dishes, but decides against it because too much friendliness with the tenants might make it harder to keep them to their contract. Through a neighbor, Ellen learns that Mr. Powers will soon be coming to trial for accidentally killing his son while swinging at his wife's lover with an axe. Ellen's response to this news is annoyance with her husband for failing to make inquiries about Mr. Powers before letting him on the place, and she is also irked that her overly easy-going mate does not press Mr. Powers for the wagon-load of firewood he owes them.

Although Ellen cannot deal with Mr. Powers on a personal basis, and he remains "Mr. Powers" throughout the story, her meditations on the landscape show that, subconsciously at least, she is aware of the injustice of the class system. She notices "that it was strange that one patch of ground should be in the deep shade and the one adjoining it in brilliant summer." At the end of the story, as she watches Mr. Powers, again from a distance, she reflects, "You could not prison light in the memory."[31] She knows that Mr. Powers faces prison, but can only articulate that knowledge by displacing it into an aesthetic observation. As she delineates Ellen's monetary and aesthetic evasions, Caroline Gordon shows how the tenant system dehumanizes landowner as well as tenant; if the owner really looked at the tenant as a person, the recognition might be unbearable.

In her second story written that fall, "The Ice House," Caroline traced the roots of the materialistic tenant system and New South to the chaos, moral as well as political, of Reconstruction.[32] Two young Southerners, Doug and Raeburn, are hired by a Yankee carpetbagger to remove the skeletons of Union soldiers from their temporary interment in an old ice house. Doug gradually loses his abhorrence of the rotting corpses and becomes interested in working hard and making as much money as he can. When he and Raeburn spy the contractor dividing the skeletons to fill more coffins and earn more money, Doug laughs admiringly. He is a harbinger of the New South, emulating the Yankee pursuit of lucre until he is oblivious of the corruption he must handle to attain it. The South has become an ice house that no longer serves its function of preserving old traditions, but is merely a dumping ground for a meaningless jumble of corrupt Yankee ways.[33]

Maxwell Perkins accepted "Mr. Powers," which appeared in *Scribner's* in November 1931. He did not take "The Ice House," however. Although his letter to her is not extant, Caroline's reply suggests the reason for his rejection: "I am sorry that you couldn't take 'The Ice House' I believe it is the best story I have ever written, structurally at least, but I can't deny that it is gruesome."[34] Caroline's usual feeling that her last piece of work was her best gave her the confidence to send it out again. If too "gruesome" for a family magazine, it could appear in a little magazine; Lincoln Kirstein took it for the *Hound & Horn* where it appeared in the summer of 1931.

Her main goal, however, was to finish her novel, *Penhally*. In March 1930, Maxwell Perkins gave her an advance of $500 after reading a draft of some early sections. He kept urging her to complete it so that he could advertise it on Scribner's next fall or spring list, but it would take Caroline another fifteen months of slow accretions, false starts, and revisions before she felt the novel was really ready to send to Perkins. By then, July 1931, she had been working on it for over three years.[35]

Although the Tates' roving, hand-to-mouth, and extremely hospitable lifestyle certainly contributed to the slow progress of the novel, there were technical and creative problems as well. *Penhally* relates the history of the Llewellyn family from just before the Civil War until the 1920s. It includes battle scenes and public events as well as domestic incidents and passions. Compressing all this into one novel, however bulky, presented a formidable technical challenge. Ford suggested that she turn it into a trilogy, probably along the lines of his own trilogy and tetralogy. Perkins discouraged this idea, citing the difficulties of marketing a trilogy, particularly selling the later books to those who feel they would then have to go back and buy the earlier ones.[36] One volume it would be, and the problem of compression remained.

Caroline cited another reason for her snail's pace. "It is the invention, not the actual writing that comes to me so slowly and it is something I can't hurry," she wrote to Perkins on November 17,

1930. Her fictions were usually closely based on facts, in this case the Meriwether family history and her own research into the Civil War. She always enjoyed the research much more than the writing; she dreaded taking that imaginative leap from fact to fiction which made writing "a torment" to her.

Her lack of confidence also delayed her progress. She was working without the presence of a mentor, except for Ford's visit in the fall of 1930, when he went over "every inch" of her manuscript.[37] She did, however, consult her peers, but she was not sufficiently sure of herself to know whether to accept or reject their advice. In March 1931 she wrote to Maxwell Perkins, "Several of my friends, and foremost among them my husband, have been objecting to the title of my book. They say that the title does not convey enough. When I say that I hope as the book goes on to invest it with meaning and point out that I have very good precedent among English novels for such a title [the name of a house] they remind me that I am neither Charles Dickens, Jane Austen, or Sir Walter Scott. They have me backed into a corner." Cowed by this unflattering comparison to a literary pantheon, she suggested "Llewellyn's Choice" as a new title. Perkins supported her original title, *Penhally,* and indeed acted as her mentor, however distant, as she worked on the novel.[38]

By the time she was completing the manuscript, she could defend her decisions against Allen's formidable critical analyses even without the advocacy of Ford or Perkins. She wrote to Sally Wood on May 30, 1931, "I went to pieces pretty badly the other night. I got frightened when Allen told me plainly that the last chapter, the climax that I had built up so fondly simply would not do. My hands got to shaking so I couldn't even hit the keys. Finally I told Allen he had to write it then if it didn't suit him. He wrote a few pages and I got interested trying to fix up what he had written—it seemed to me so impossible—that I worked out of the fit."[39] She was no longer the young woman who destroyed her first novel because Allen's expression had told her it was no good.

Although she "worked out of" that "fit," she felt she needed to steel herself for the book's reception. She received her first copies of *Penhally* in the middle of September and commented to Perkins, "It looks very well, I think, and is beautifully proof read." This low-key reaction concealed her actual anxiety. She wrote to

Sally Wood, "Dickson-Sadler's the drug store is going to have a window display of my book with the author autographing—if the kin don't come and at least stand around in the attitude of people about to buy books I shall be peeved. I'm nervous about it, fearing it will be like getting ready to be married and no groom turning up."[40] The occasion was a success and Caroline's debut as an author was a local triumph. Her cousin Danforth Ross remembered seeing her at a dance at Dunbar cave shortly after publication. "She was wearing a long black dress. . . . Her expression struck me as arrogant, as if she were carried away by the fact that she had just published a novel. I am sure that this was a big moment for her."[41] What Ross perceived as arrogance was probably the mask-like demeanor she could assume when trying to hide her fears and appear less vulnerable.

The "kin" did support the book, probably because it is better to claim an author in the family and be put into a book, however unflatteringly, than to remain without such immortalization. Much to Caroline's surprise, her grandmother Miss Carrie took a "conventional pride in having a granddaughter who had written a book." Caroline's cousin, Norma H. Struss, recalled that *Penhally* "was supposed by many to be her portrayal of the family over the years which naturally produced much comment. She seeming to feel they were inclined to not use their potential and see themselves as being far better than they were."[42] Since that was also James Gordon's opinion of the Meriwethers, he liked the novel so much that he read it twice "and explains it amuses him because he knows all those people so well."[43]

The response to *Penhally* outside the environs of Clarksville was less of a triumph. It did not sell well, not even making enough for any profit above the advances she had received. She found the reviews "disheartening" and felt that the reviewers did not understand what she was attempting in the novel.[44] Even before publication she had foreseen one reason for the reviewers' incomprehension; she wrote to Josephine Herbst that the book was "*too* short, too compressed. . . . I am not sure that the average intelligent reader will ever see what I've tried to do." *Penhally* is a difficult novel to read, but its compactness is not the only barrier challenging the reader.

A bald summary of *Penhally*'s plot seems to characterize it as a fictionalized treatment of the Agrarian manifesto. In 1826 two half brothers, Nicholas and Ralph Llewellyn, quarrel when the younger, Ralph, wants to divide the Penhally tract to have a house of his own. Nicholas believes in the pre-Jeffersonian practices of entail and the communal family and refuses to divide the property. Ralph does not press the issue out of deference to his aged stepmother, but moves to a nearby farm, Mayfield, and raises race horses. At the outbreak of the Civil War, the brothers again differ. Ralph beggars himself for the Confederacy and loses his only son in battle. Nicholas buries his gold instead of buying Confederate dollars and manages to keep Penhally together. His heir, John, a cousin, returns from the war and marries Ralph's daughter Lucy.

As John and Lucy face the problems of reconstruction their marriage disintegrates. Their only child, Frank, makes an unfortunate marriage and commits suicide. His sons, Nicholas and Chance, are returned to Penhally to be raised by their grandparents, Lucy and John. Although he later inherits Penhally, Nicholas rejects the life-giving soil for the sterile existence of a banker in town. Chance, a born farmer, manages Penhally for his brother. Nicholas agrees to an advantageous offer to sell Penhally to a fancy hunt club. Chance shoots Nicholas. The pastoral order of antebellum days crumbles into commercialization and fratricide, seemingly according to the Agrarian scheme.

Penhally has usually been read as just such an elegy for antebellum days destroyed by the forces of history, probably because Caroline was the wife of Allen Tate and closely associated with the Agrarians. The forces of history are not the villains of her piece, however. When she studied with Professor Gay at Bethany College, Caroline had immersed herself in what are, after all, Greek family tragedies, and like them, *Penhally* emphasizes the importance of bloodlines. Heredity is destiny and much of the novel hinges on the contamination of "good" blood by intermarriage with "bad" blood and the various permutations of inherited traits.

These crucial facts are what Caroline's modernist method of presentation tends to obscure. Caroline is attempting to relate a multigenerational family saga with a plethora of cousins, neighbors, and plantations with the same or similar names. Such a broad and crowded canvas ordinarily would be presented in as traditional and straightforward a manner as possible, as in the novels of Galsworthy, but Caroline chose instead the Jamesian methods she had learned from Ford.

The story is told from the points of view of a score of characters. The author does not intrude to convey information; we are expected to glean the facts from the consciousness of the various characters. Important bits of information slip casually down the stream of consciousness as the reader attempts to keep the tribe of Llewellyn and their retainers straight. The reader's need for the facts of the complex family saga is in violent conflict with the author's manner of relating them, or, as Robert Penn Warren puts it, Gordon's "extensive" subject is in conflict with her "intensive" treatment.[45]

Even though she often felt "sick" over *Penhally*'s flaws, Caroline had the satisfaction of finally completing and publishing a full-length work. She was also heartened by letters from friends and fellow authors, such as Josephine Herbst and Stark Young, who appreciated the novel's densely interwoven texture of imagery and setting.[46]

Her new status as a published novelist was not readily recognized by all, however, and she was not invited to the Southern Writers Conference that was to be held in Charlottesville, Virginia, in October 1931. Her letters of the time contain comic references to the fact she was "uninvited" and would have to attend as a "stowaway," but might prefer to stay home since her dog was about to have puppies.[47] Allen, of course, had been invited. Caroline's chagrin was the beginning of a pattern that would poison their marriage; she felt that her work remained comparatively unnoticed, overshadowed by Allen's reputation.

In the end, Caroline did attend the conference as a spouse, but injury was added to insult, as Allen wrote to the poet Virginia Tunstall. After two days of silent and steady drinking, William Faulkner accidentally spit up some of his drink on Caroline, and on the new dress she had labored over for several weeks; worse still, he did not seem sorry.[48] It is hardly surprising that Caroline did not like Faulkner personally, though she admired his work.

Caroline was somewhat consoled by the presence of a fellow "stowaway," Andrew Lytle, who shared her gloomy feelings about the conference. She wrote, "Virginia is terrible. The buzzards hang around all day over the fields—scenting the odor of decay Andrew says."[49] Andrew also supplied a lighter note by becoming infatuated with one of Caroline's Virginia cousins, Sarah Lindsay Patton. Caroline wrote to Sally Wood that Patty, as she was called, "works on the garden barefooted—the negro house boy said 'I can always tell when Miss Patty's beaus is getting serious. They takes off their shoes.' Andrew was barefooted thirty minutes after we arrived."[50] Although the romance did not prosper, it served as the basis for a lot of good-natured teasing from the Agrarian brethren, particularly since before his marriage Allen had been very taken with Patty's older sister Alice.

Although Caroline could joke about her lack of status at the Southern Writers Conference, she was becoming increasingly conscious of the difficulties of a woman writer. In August, she had written to Perkins, "It is, certainly, much harder for a woman to write than it is for a man. I am in a panic half the time fearing something will happen to prevent me from writing. But I am very fierce about it, I assure you. And we have a wonderful servant [Beatrice] who does everything she can to help me get on with my writing."[51] The Tates' financial exigencies were worsened by *Penhally*'s failure to sell so that Caroline's work was once more delayed by domesticity. She expressed her frustration to Josephine Herbst: "Perkins urges me not to be discouraged but of course I am. I have another one all ready to write but can't get at it on account of being so broke. I had to let Beatrice go—couldn't pay her."

The novel Caroline had "all ready to write" was the as yet untitled *The Garden of Adonis*. She had touched on a similar subject in "Mr. Powers" in the fall of 1930 and had clarified her conception of the novel by the spring of 1931: "Two families, white and poor white, living on the same farm. The situation seen through first the eyes of one and then of the other. Each regards the other as his natural enemy."[52] As she told Perkins, "Its framework would be the tenant system which has grown up in our country since the Civil War. It is a kind of bastard feudalism, highly organized in a way, yet yielding as many emotional complications as an unhappy marriage. I have been fascinated ever since I was a small child by its complicated workings."[53] Due to a paralyzing array of personal and creative difficulties, *The Garden of Adonis* would not be completed and published until 1937.

Although she was unable to make progress on *The Garden of Adonis,* she had managed to complete a long story, "The Captive," based on a historical work by William Elsey Connelly, *The Founding of Harman's Station with an Account of the Indian Captivity of Mrs. Jenney Wiley (1910).*[54] "The Captive" was intended to be the first piece in a short story sequence, arranged chronologically, about early Tennessee. Caroline described it to Perkins in July 1931: "The pioneer, I suppose, would be the hero—or the country. I want to fit them [the stories] all into the same background. I have half a dozen stories in mind now, each a sort of adventure in itself, but dealing with a certain phase of the life. . . . I want to write them as one book, plan the detail, in fact, pretty much like the chapters of a novel." Perkins discouraged her from pursuing this plan because of the difficulty of selling collections of stories, but Caroline would use it, somewhat modified, in her first collection of short stories, *The Forest of the South* (1945).[55]

In "The Captive," Caroline transformed Connelly's dry historical account into a suspenseful and compelling story of courage and endurance. She streamlined the narrative and quickened its pace by omitting some of Mrs. Wiley's trials and compressing the period of her captivity. Caroline's most effective method of dramatizing Mrs. Wiley's ordeal, however, was casting it in the first person. The reader experiences events with Jinny Wiley and shares her fear and uncertainty as well as her eventual triumphal escape.

In an interview in 1966, Caroline Gordon emphasized Jinny Wiley's heroism by contrasting her with contemporary women. "I would have been dead nuts by this time, and I think most modern women. She is of heroic stature but still she is a human being. . . . I think she's got about forty times as much moral fibre as almost any woman I know."[56] When she wrote the story in the fall of 1931, Caroline was lost in a forest of self-doubt and may have felt some measure of relief in speaking with the voice of a woman who confronted the dangers of a literal forest and survived. By contrast with the contemporary section of *Penhally,* the story also reflects Caroline's sense of the modern world as fallen and diminished; its denizens no longer have the opportunity or capacity for heroic action.

The Tates' finances and Caroline's ego were to be further tried. Since they were so pressed for cash, Caroline wired Perkins for an immediate decision on "The Captive." In a telegram dated January 13, 1932, he replied "Magazine does not think Captive has enough regular story interest to publish as a long story." As he later explained in a letter, he meant that this first version of the story was too much like a historical chronicle and Mrs. Wiley did not emerge as a "real character," with psychological depth, until the second half of the story.[57] Caroline thought that his telegram meant that he did not believe the story had an exciting plot. She wrote to Robert Penn Warren that "My rage was so great that I couldn't use my mother tongue properly so Allen wrote for me, a deadly letter telling him what was what and asking him what the hell and so on."[58] Allen did not merely write in her behalf; he wrote a letter that she signed as her own.

Caroline's description of Allen's letter is a good one for it sounds just like Allen in a mood of icy disdain and not a bit like Caroline's more hesitant and accommodating letters to Perkins. Despite the fact that she professed herself tongue-tied with anger, it is likely that her cultural conditioning as a woman, a Southern woman at that, made her as yet incapable of writing an assertive letter that accused her publisher of forcing her to do "hack-work" to support herself and proclaimed the merits of her work. She may have been trying to have it both ways: although she wanted to be invited to a writers' conference on her own merits, she still depended on Allen for the less pleasant aspects of literary survival.

The letter is also interesting as Allen's assessment of Caroline's work and his role in promoting it. In her name, he wrote, "I am not a talented amateur; I am as mature as I shall ever be. . . . I am not only at my best right now; I am at my best because I am one of the few writers in this generation who have something to write about. . . . But instead of publishing me now, Scribner's Magazine will be publishing my imitators then [in five or ten years], and then will probably use, unconsciously, the state of mind my work has induced to resist some other new talent." Allen obviously valued Caroline's work for its technical innovations and its subject matter, but he was also getting in a blast at what he considered the artistic philistinism of the literary establishment. In the future, he would occasionally intervene with publishers for Caroline, and although his interactions were often astute, they made it much harder for Caroline to act for herself when he was no longer there to act for her.

In this case, Perkins smoothed Caroline's ruffled feelings by a well-timed reference to his hopes, unrealized, that *Penhally* would win the Pulitzer Prize. He also carefully explained how he thought Caroline could improve "The Captive."[59] Caroline wrote to Red Warren that Perkins' letter "was masterly and had me roped and tied in two paragraphs and then went on to do his will. That Perkins is a smooth article and there is no use fooling with him." She decided to revise the story to give Mrs. Wiley "a little personality."[60] The *Hound & Horn* published "The Captive" in their final issue of 1932.

The Tates' troubles in the winter and spring of 1932 were compounded by illness. Both Caroline and Allen had what Allen called "influenza of the intestines" in late January, but while Allen recovered, Caroline continued to experience symptoms that she compared to labor pains.[61] She had a curetage, possibly an abortion, in March, but her improvement was only temporary.[62] The family doctor speculated that she might have a cyst on one of her ovaries, but finally decided that she had appendicitis.[63] On April 20 she had her appendix removed at the Clarksville Hospital and spent the next two months recuperating.[64] Although her poor health prevented her from making any progress on her novel, she toyed intermittently with writing a fairy story for children, a curious birth following her recent labor pains. "My idea," she wrote Sally Wood, "was that I might avoid starving if I could only learn to write a juvenile, say one a year."[65] This hope remained unfulfilled, but Caroline continued to consider writing fairy stories and mysteries between her "serious" novels.

Caroline was sustained through this period of illness, inaction, and insolvency by a delightful prospect for the summer. In mid-March she learned that she had received a Guggenheim award of two thousand dollars. The Tates decided to leave their troubles behind them and return to France for about six months. Effective April 15, they rented Benfolly to some local newlyweds and moved in with Miss Carrie at Merimont since they did not plan to embark until the end of July. Allen busied himself with fixing locks and screens while Caroline luxuriated in her convalescence and her freedom from housekeeping. She wrote to Sally Wood, "The dirt which appalls at first comes to seem an advantage. One simply picks one's way about over old stacks of magazines—and it is a relief never to have to think of cleaning up. I am really quite addicted to the place—I much prefer it to Benfolly."[66]

After some vacillation, the Tates decided that Nancy would not accompany them to France this time. Caroline's Aunt Piedie, Margaret Meriwether Campbell, offered to take her into her home in Chattanooga for the year. Almost seven now, Nancy could attend school there and have Piedie's eleven-year-old daughter for a companion. Caroline continued to express misgivings throughout the trip, but reasoned, "I would feel as safe about her as I possibly could. Still it would put the ocean between us. On the other hand I could do a hell of a lot more work." Caroline reassured herself further by sewing a year's worth of clothes for Nancy.[67]

The Tates had a full itinerary before they even left the pier in New York on July 21. They planned to take their Model A Ford to Europe with them and so used it in their travels. First they took Nancy to Chattanooga which, wrote Caroline, "exhausted us to begin with. Leaping from Lookout to Signal [Mountains] for lunch and then back to Lookout, say to swim in the Fairyland Club." Their complex peregrinations also included a trip to Monteagle, Tennessee, where they visited with Andrew Lytle, his sister Polly, and the poet Phelps Putnam in the Lytle cabin. Allen, Caroline, and Andrew gave lectures and readings at Monteagle. The Tates' talks were uneventful, but Lytle caused a stir, as Caroline related to Red Warren: "Then Andrew came on and gave them resumes of Broadway plays. The plays, unfortunately, all seemed to deal with unsatisfactory marital relations . . . and once when hard pressed he simply waves his hand 'Well, what I mean is his pappy warn't his daddy nohow.' One old lady turned to another and said 'Blasphemous,' and one or two got up and walked out—a strange effect as if boulders were turning over in their sleep."[68]

The whirl of activity continued when the Tates arrived in New York where they stayed in Malcolm Cowley's apartment. They were besieged by old friends and literary contacts. Allen and Caroline had a lunch with Maxwell Perkins at which the conversation "never got off the Civil War." Caroline pronounced Perkins' views "very sound" and Perkins himself "the nicest man," adding, "You cannot understand how he can function as a publisher."

One evening the Tates had dinner in a speakeasy with their old friend from the Tory Valley, William Slater Brown. There they encountered Edmund Wilson who "lectured" them throughout dinner, according to Caroline. After dinner Wilson invited them to his apartment but left them alone while he went upstairs to assist a man who was working on a concordance. The surreal quality of the evening continued with a drunken taxi ride in which Caroline and Bill Brown sang and put their feet up on the seat. When the driver asked if they wanted a hotel, the others "repudiated the idea at once but Edmund considered it as he does all ideas and then said severely, 'No, that is not what we want.' "

Bernard Bandler, one of the editors of the *Hound & Horn,* gave the Tates a bon voyage party. Caroline remembered "rushing up to Bernard's apartment just in time to tune in on Allen on the radio. It is a devil box, the radio—so strange to hear Allen's voice coming out in those poems I know so well." The exhausted Tates departed the next morning, July 21, on the *Stuttgart* for what must have been a welcome "dull crossing," tourist third.

If the crossing was dull, it certainly wasn't solitary. Typically, the Tates were travelling with five companions. Allen's fellow Agrarian, Lyle Lanier of Nashville, and his wife "Chink" would stay with the Tates in Paris until September. Also included was Caroline's old friend and correspondent Sally Wood who would remain with them until the end of November. Caroline also invited her first cousin and childhood companion, "Manny," Marion Douglas Meriwether because "she hasn't had a vacation in God knows when and leads the most desolate life in that girls' school and hates the whole thing so thoroughly I jumped at the idea of her going."[69] Manny stayed with the Tates until school resumed in the fall. Finally, there was the sixteen-year-old Dorothy Ann Ross, one of Caroline's distant cousins, who needed a chaperon on her way to school in Switzerland.

The Tates and entourage arrived in Paris and settled down in Ford's apartment at 32 rue Vaugirard since Ford was in Italy. It was not the apartment they had borrowed on their last trip, but the nearby one formerly occupied by Stella Bowen. More recently, it had sheltered Katherine Anne Porter before she left for Germany. Caroline learned from the concierge that "K. A. is not a good menagere but I told her she was a fine writer. Still Mme Jeanne thinks she shouldn't have lost Ford's hot water heater."[70] Despite the lost heater, Caroline declared the apartment "very gemutlich" with "enough familiar objects to make us feel at home."[71] Understandably, considering the size of the household, Allen soon left for Orgeval to visit John Peale Bishop.

While the other members of the party were out shopping and sightseeing, Caroline usually remained in the apartment. She was still suffering from soreness in her side which made her reluctant to climb the six flights of stairs to the apartment. For this reason and others, Paris seemed to have lost some of its enchantment. Now, she wrote to Malcolm and Muriel Cowley, Paris "roars" because "they have doubled the number of autobusses. . . . Also I grieve to report . . . that every restaurant we used to go to is just a little bit higher." Furthermore, although August, it was "as cold as the devil."[72]

Despite the roaring below, Caroline was trying to work on a short story, "Tom Rivers," based on her Uncle Rob's adventures in Texas as a young man. Tom Rivers had left home for Texas to flee the restraints of a matriarchal family. The last straw had been his girlfriend's demand that he quit drinking. He is joined there by his younger cousin, Lew Allard, the narrator of the tale. Lew characterizes Tom as "utterly fearless" and in constant motion, two characteristics that make him unsuited for domestic life. Lew returns to the family after the adventures with Tom that constitute the burden of the story. Although Tom has physically escaped the matriarchy, he remains enmeshed in the web of family memory when his deeds are discussed under the enveloping bulwark of the sugar tree, suggestive of the family tree.[73] In "Tom Rivers," Caroline treats the conflict between freedom-loving males and the demands of the family, characterized as female, which she would develop at length in her next published novel, *Aleck Maury, Sportsman.*

After rejections from *Scribner's* and the *Virginia Quarterly Review,* Caroline sent "Tom Rivers" to the *Yale Review.* They wanted the story, but felt it needed some revisions, and the Tates repeated the scenario they had enacted with Perkins over "The Captive." An editor at the *Yale Review* wrote that "she had taken the liberty of revising the manuscript in three or four places, inserting quite a lot of her own writing. I was just sitting down to thank her for doing it when Allen glanced at the letter and said that he thought it was unwarrantable insolence."[74] In this instance, Caroline replied herself, diplomatically reinstating her work, but offering to cut some passages later in the story. Her tactful approach was successful and the story appeared in the October 1933 issue.

In September, after the departure of Marion Meriwether and the Laniers, the Tates and Sally Wood headed south for Toulon and Ford Madox Ford. With his latest and last wife, the painter Janice Biala, Ford occupied the Villa Paul. He had obtained the nearby Villa des Hortensias for the Tates and Sally Wood. Caroline had looked forward to being with Ford again. As she planned the trip, she had written to Sally Wood, "I really like being in his vicinity. To have even one person planning things so you can get your work done helps a lot. And his mild tyrannies will be exerted over Janice this time, let us hope." She also joked to Wood about Ford's propensity to fall in love with any ladies in the vicinity: "you're not likely to fall in love with Ford and I for one will be rather pleased if he falls in love with you. You've stood exposure to Allen very well."[75]

Caroline's high hopes for the visit were disappointed, largely because Allen believed he was out of his element. While Ford and Caroline had long discussions about the novel, Allen spent some time talking to Sally Wood, some time in a cafe reading Virgil, and a great deal of time worrying over his lack of progress on his biography of Robert E. Lee.[76] Although he and Caroline attributed his trouble to the absence of suitable libraries in Toulon, he had really lost interest in the subject and was even beginning to question the efficacy of his Agrarian beliefs. Later that year he wrote to Donald Davidson, "The trouble with our agrarianism is not that we don't believe in it enough to make sacrifices; it is rather that we don't believe in it in the way that demands sacrifices. In other words not one of us has a religion that any of the others can understand."[77]

Although Allen felt that he was wasting his time in Toulon, he was actually writing some fine poetry. His reading of Virgil inspired "Aeneas at Washington" and his great poem "The Mediterranean." A picnic at Cassis in October, before Sally Wood left the group and the Tates returned to Paris, furnished the inspiration for "The Mediterranean." The poet longs to turn back to the days of Aeneas, not the Confederacy:

> Let us lie down once more by the breathing side
> Of Ocean, where our live forefathers sleep
> As if the Known Sea were still a month wide—

He questions the virtue of the pioneering spirit that settled America: "We've cracked the hemisphere with careless hand!" After the movement "Westward, westward," we are left in a land of exhausted and sinister fecundity, suggestive of the South: "the tired land where tasseling corn,/Fat beans, grapes sweeter than muscadine/Rot on the vine: in that land we were born."[78] Tate's rejection of provincial concerns and longing to burst the bounds of temporality would provide part of the impetus for his conversion to Catholicism almost two decades later.

Perhaps in the spirit of "if you can't beat 'em, join 'em," Allen also began to experiment with fiction. He wrote an oblique Jamesian short story, "The Immortal Woman." He also explored the themes of "The Mediterranean" in prose as he began *Ancestors in Exile* in which he wanted to demonstrate the way the pioneering spirit undercut the values of a settled traditional society and led to the chaos of modernity. Although he did not complete this book, both it and "The Immortal Woman" served as Ur-versions of his only novel, *The Fathers* (1938).

Caroline, too, managed to accomplish some work. She produced three more chapters of her novel, *The Garden of Adonis,* and continued to write "Tom Rivers." Although the south of France did not immediately inspire a work, as it had in Allen's case, years later she would use the picnic at Cassis in her novel *The Strange Children* (1954) and the Villa de Hortensias as the setting of her short story, "The Olive Garden" (1945). Allen's meditations on the pioneer spirit would later inspire her novel *Green Centuries* (1941).

In the hope that Allen could make some progress on Lee, the Tates headed for Paris in mid-November 1932. It was a foggy journey, but they enjoyed stops at the Roman tombs at Arles and the Palace of the Popes at Avignon. When they arrived in Paris, they needed to find a place to live since Ford was returning to his apartment in January. At 32 Denfert-Rochereau they rented a studio consisting of one long room, a small kitchen, and a balcony bedroom; since the amenities did not include running water, it is not surprising that Caroline described it as "the dirtiest place . . . I ever saw in my whole life."[79]

Although the Tates believed themselves unproductive amid the beauties of Toulon, they seemed inspired by the primitive discomforts of a Paris studio. On November 23, 1932, Caroline wrote to Sally Wood:

> We are both working like hell, have settled down into one of those routines in which people do produce books. It's funny. God—or the devil somehow creates these islands about once a year for us, the only time you seem to get anything done. There is not one soul in Paris I feel even a passing interest in. Allen works here at home in the morning. We have lunch. I go out to the market—a swell market around the corner—come home, sleep till two o'clock, get up and go at it again, till five when I walk briskly around the court eight or ten times, come in and get supper. After supper we work some more and so to bed. If it will only last.[80]

It did not last, and the Tates decided to move to the Hotel Fleurus so that Caroline would be spared housework.

An influx of friends and acquaintances over the holidays further disturbed their idyll. The Tates spent Thanksgiving with fellow American Walter Lowenfels, Marxist poet and editor, and his wife Lillian who cooked the turkey. Katherine Anne Porter was back in Paris and lived near the Tates. To Janet Lewis, novelist and wife of poet Yvor Winters, Caroline reported that Katherine Anne "is in love—and it looks very serious this time. . . . He is very nice, from all reports and very devoted to K.A. and properly appreciative of her talents. I kept wondering why she looked so different and finally realized it was because she was really happy for the first time in her life."[81] The source of this change was diplomat Eugene Pressly, and the Tates had Christmas dinner at the Couchon au Lait with the couple. From Stella Bowen, the Tates received the portrait of the three of them that she had painted during their previous stay in Paris. Nancy was not forgotten either, since her parents had sent her a rocking horse and rubber doll well in advance.

The Fords were arriving in mid-January, so Caroline and Katherine Anne busied themselves getting the apartment ready. They met the Fords' train and took them to lunch. Unfortunately, Ford was in a troubled and troubling mood that winter, as Caroline explained to Sally Wood. "Ford, faced with privation, has got very hoity toity—poor devil. Janice told K. A. he walks the floor at night and pictures Julie [his daughter] starving. It is at times like this when it is very difficult for his friends to rally round him."[82]

Ford was not the only difficult older writer Caroline encountered: "we were taking a turn in the gardens the other day and I saw streaking towards us an oversized white French poodle and one of those Mexican toy terriers—chi—what do you call them? 'My God, look at that,' said I, meaning the remarkable pair of dogs. Allen, having an eye more for the ladies said, 'Yes, it's Gertrude Stein' and so it was. She stopped and had a chat. . . . Gertrude was surprised that it was me and not Allen who had the Guggenheim this time. 'You?' she says 'And what can you do?' 'I'm trying to write a novel' I rejoins meekly."[83] Caroline's "meekness" undoubtedly concealed her glee at showing up Stein who had always relegated Caroline to the corner of her *salon* she reserved for female nonentities.

Caroline may have been "trying to write a novel," but she was not having much success. After the New Year she was felled by "the grippe" and lost a month's work, making her "very bitter." This "January collapse" became annual as did her self-reproach at her inability to work. To make matters worse, Katherine Anne Porter told her that the manuscript of *The Garden of Adonis* "sounded like somebody trying to write like" Caroline. When she managed to complete a short story, "Old Red,"[84] Sally Wood, Ford, and John Peale Bishop informed her that it did not have enough action. Caroline humbly reflected that "I was really more interested in rendering the character of the man than I was in the action of the story and that always betrays you."[85]

Ironically, "Old Red" is usually considered her best story, and is certainly the most frequently anthologized and analyzed. In "Old Red," Caroline introduces her best and best-loved fictional character, Aleck Maury, who was modelled on her father. Caroline discovered the "germ" of her story during her last stay at Merimont when James Gordon took Allen fishing with him instead of attending the funeral of an elderly female relation.[86]

"Old Red" focuses on the dichotomy between masculine and feminine conceptions of time. Time seems to stand still for the women of "Merry Point": the furniture, the activities, even the slant of sunlight remain as Aleck Maury remembered them. The women still act as if time were their ally and they had all the time in the world. For the men, in contrast, time is an adversary to be pursued and caught, the ultimate masculine sport. Doing is much more valuable than being, so accomplishments are the measure of success. Despite the pleasant weather, Aleck's son-in-law works compulsively on his essay. Aleck himself evaluates his time in terms of what he has learned about fishing and what he has caught.

Caroline herself lived in constant tension between what she regarded as the masculine accomplishments of publications and the endless process of feminine domesticity. In one of her most beautifully rendered passages, through Aleck Maury she envisions the gender-differentiated conceptions of time that are central to her works and to her life.

> *Time,* he thought, *time!* [The women] were always mouthing the word, and what did they know about it? He saw time suddenly, a dull, leaden-colored fabric depending from the old lady's hands, from the hands of all of them, a blanket that they pulled about between them, now here, now there, trying to cover up their nakedness. Or they would cast it on the ground and creep in among the folds, finding one day a little more tightly rolled than another, but all of it everywhere the same dull gray substance. But time was a banner that whipped before him always in the wind! He stood on tiptoe to catch at the bright folds, to strain them to his bosom. They were bright and glittering. But they whipped by so fast and were whipping always ever faster.[87]

Caroline's ambivalence is further demonstrated when Aleck Maury's daughter Sarah, the Caroline-figure, wants him to attend an old kinswoman's funeral. He identifies with the wily, hunted fox, Old Red, suggesting that Sarah and the other women are the potential captors. Escape is necessary so he leaves the feminine preserve of Merry Point to continue his own hunt for time as he pursues sport throughout the South.

In the winter of 1933, Caroline did not know that her story would be a success, and it simply added to her depression. Allen was finding Paris dull. Caroline was once again tired of the city and longing to return to Benfolly's bucolic delights. Nancy was also weighing heavily on her conscience, particularly after she learned Nancy had the grippe in Chattanooga. Caroline vowed to Sally Wood, "I don't know how I would ever have done any work if I'd had her along but I am never going to put the ocean between us again, work or no work. It's been better for her, but it's hell on me. Every night when I go to bed I am convinced that she will die before we get back."[88] When Caroline learned that her Guggenheim would not be renewed, the Tates were glad to head for home. Their dismal stay in Paris ended on a happy note, however, with a dinner given by Sylvia Beach and Adrienne Monnier that Allen remembered as "delightful" for "the warmth and the lively talk."[89]

Since Benfolly was rented until August, the Tates were in no hurry to arrive in Clarksville. They visited Phelps Putnam in Maryland for two weeks, and then proceeded to their many friends in Nashville for another fortnight. By April, they were back at Merimont, much to Caroline's satisfaction. As she had written to Sally Wood, "We can live at Merry Mont for practically nothing and the money we'd be spending here can go to pay debts." Nancy was to remain in Chattanooga to finish the school year, so Caroline would be free of childcare as well as running a household. She also reasoned that "if this inability to write continues there will be other things to do" at Merimont, and do them she did.[90] She reported to Josephine Herbst that she worked on the garden and made twenty-one pints of strawberry jam because she was determined to preserve enough to last through the winter at Benfolly. As she wrote to Malcolm Cowley, "Most people would want to cut their throats after a few weeks of Merry Mont but I really enjoy it."

Allen, unfortunately, was more of the throat-cutting school; although Merimont did not drive him to suicide, it did make him exceedingly restless. The Meriwethers did not regard writing as work so they thought nothing of interrupting him to serve as a driver. He was even required to drive old Uncle Nick, Miss Carrie's black factotum, out to the sheep.[91] Allen was essentially an urban person who needed an intellectual society the Meriwethers could not provide. Luckily, Malcolm Cowley arrived in May to board with the Henry Meriwethers at nearby Cloverlands while he worked on *Exile's Return*.

Cowley remembered the insular nature of the family farms: "We were fourteen miles from Clarksville, three miles from a telephone, and a mile from the mailbox at the end of the lane, which last was our principal connection with modern America. There was no radio in the house, and the only newspaper, which arrived a day late, was the Clarksville *Leaf-Chronicle,* devoted chiefly to local doings and the price at auction of dark-fired leaf tobacco."[92] After the morning stint at the typewriters, the three authors were dependent on each other for companionship during the hot afternoons of swimming and fishing. Sometimes Caroline's Uncle Rob, whom Cowley remembers as "an old shiftless drunk" but "really a very charming man" would take them over to Hopkinsville to buy liquor.[93]

The threesome was augmented by one of Caroline's cousins, dark, attractive Marion Henry. Before the Tates had left for Paris, Caroline empathized with Marion's isolation and inability to find meaningful work within the Meriwether connection. She encouraged her to apply for a job in a Nashville library. Her application was unsuccessful, as Caroline wrote to Sally Wood, because "her attitude and address weren't right, but that is mostly just being brought up in this out of the way part of the world where you have to keep saying something pleasant whether you feel like talking or not. She read the proof on my novel [*Penhally*] in a most business like way and caught all sorts of tricky little errors. She is probably more capable and intelligent than most people making a living."[94]

Marion's lack of meaningful employment ended in a tragic way when her father became incapacitated by a stroke and she assumed the management of his farms, over one thousand acres.[95] Cowley remembered her as an obsessive farmer.

> Once, she said, blinking her dark eyes, she had wanted to *create*—I winced at the word—but now all she wanted to do was get land, land, land, more hundreds and thousands of acres, all the farms in the neighborhood. She closed her fist as if she were squeezing the farms together. "I want all the land my grandfather owned," she said. "That was more than five thousand acres, and it's all I want in the world." "Who'll get it when you're dead?" "I don't care. I'm all there is." Her eyes frightened me as they glowed in the dusk. "It all comes down to me. I don't care what happens to it after I'm gone."[96]

Her passion for land went hand in hand with her staunch devotion to the South. She would not enter a Union cemetery at Fort Donelson, declaring, "I wouldn't be caught alive or dead in a Yankee burial ground."[97]

Marion Henry was much as Caroline had been nine years ago: an intense, attractive, dark young woman with literary, Southern, and agrarian inclinations. History repeated itself: Allen fell in love with her and a passionate affair ensued.

The romance was short-lived, ending before Cowley left in mid-July, but it was devastating to both Allen and Caroline. Allen told Cowley, "it's broken up now and I'm cut to pieces." He left the area without telling anyone where he was going and did not return for several weeks. Caroline would not speak of the situation at all. Cowley recalled, "for some weeks Caroline and I were left alone together. Now, don't imagine any hoopla because Caroline was one of the chastest women I have ever met, not that I tested her chastity."[98] Cowley's opinion of Caroline's chastity was apparently accurate for however she later revenged herself on Allen for his affairs, it was not by having them herself. On Allen's return, the Tates set off for Guntersville, Alabama, to stay at the farm of Andrew Lytle's father. The urgency of their need to leave the scene of the *crime passionel* is demonstrated by the fact that they left behind their friend and "guest" Malcolm Cowley who was to remain at Cloverlands for another six weeks.

In a letter she wrote to Sally Wood from Guntersville, Caroline, like Emily Dickinson, seemed to want to "tell the truth but tell it slant." She informs her friend that they left Merimont because she was suffering from a "minor nervous collapse," which she attributes to a number of causes. She explains that Merimont was "sinister," and that going there "was a mistake, I reckon. It is just too hectic and the place has a very bad influence on me and through me on Allen. The family responsibilities just kept getting heavier and heavier." She tenders another explanation: "It's that wretched back of mine. I will confess to you what I've concealed from the family so far. I hurt the damn thing pulling up some enormous weeds in the garden. The ache lurked about for a while then transferred itself to my knee just as it did the time I had the three months' breakdown in Chattanooga [when she was writing her first novel in her aunt's attic]." She concludes the letter with yet another explanation, this time a joking one: "I learned to drive. Think of that. Maybe that gave me the nervous breakdown."[99]

All these explanations for her state of mind have a common theme: she felt in some way to blame for her own pain; the "enormous weeds" were not entirely responsible, but her decision to root them up. Perhaps she had attacked Allen's faults, his weeds, too vigorously. Maybe she had neglected him for "family responsi-bilities." Perhaps since the publication of her book, she had be-come too independent; learning to drive or write meant she would not need to rely on Allen as the one driver, writer, and breadwin-ner in the family. Since her family was part of the source of her pain, she could not "confess" to them, but found some relief in an oblique revelation to a distant friend.

Caroline also found some relief in her work, even though she felt unable to continue *The Garden of Adonis*. It would be interest-ing to know if the infidelity subplots of that novel were added after Allen's affair, but it is certain that most of Caroline's remaining novels contain women tormented by the knowledge of their hus-band's adultery. She could not tackle this subject yet, however. As she later wrote, "If I was to work at all I had to work through somebody else's mind as my own had been rendered unendurable to me."[100] She chose her father's mind, the masculine point of view she had handled so successfully in "Old Red." Thus *Aleck Maury, Sportsman* was conceived; her most sunny and approachable book would emerge from her period of anguish.

Caroline's intent in this novel is best expressed by its original title, *The Life and Passion of Aleck Maury*. Through Aleck Maury she wanted to prove that a person could pursue his passion without sacrificing the other aspects of a normal life, which was, of course, the balancing act she was attempting in her own life. As she ex-plained to Maxwell Perkins, "I have tried to take care of the con-flict that would inevitably arise out of his marriage by making Maury evince his peculiar talent for making his life into a satisfac-tory pattern. Mary [later Molly] is a high spirited woman—and no woman likes to have her husband devote himself wholly to angling. Maury circumvents her not by conflict or argument but by having for her the sort of passionate devotion that would appease any woman. Similarly for his relationship with his children. He is a good father." Her plan for the book, however, and the completed novel itself, end not with a "satisfactory pattern" or balance, but with "the scene in which, not without pain, he sacrifices the last ties of family affection in order to devote himself to angling."

In a strange confluence of life and art, Caroline's father was to "sacrifice the last ties of family affection" by refusing to take the time to work on the book with Caroline. He wrote to her:

The Prospectus of the Book sounds very intriguing, but I can not do it now: for years I served a bitter Bondage in the School room: and then a worse one in the Pulpit: now *I am free* and do not intend to get entangled again in the Yoke of Bondage. . . . You can not conceive, for you have not been thro it, the joy of not having to do any thing yoú dont want to do! The few years left I intend to revel in that joy, and any set Task would be a Fly in the Ointment—I hate to refuse you, but I just could not do it: I could not give a single morning to it, when the Wonderful "Out doors" is calling me. I wish I had been born 150 years ago & could have lived the glorious life of a Pioneer.[101]

This passage not only shows how close Aleck Maury is to his prototype, but how much Caroline resembled her progenitor, both in inclinations and gifts.

Since her quarry refused to become "entangled" in the net of Benfolly, in early October 1933 Caroline left to seek him at his current lair, the Hill Side Inn in Walling, Tennessee. As they sat on the verandah overlooking the Caney Fork River, she set her trap. "I knew that if I once got him to start talking, my story was half written. I therefore placed my typewriter on a small table near the chair in which he sat. . . . As time went on, he warmed to the sound of his own voice. One morning at two o'clock he broke the silence of the lodge by shouting, 'Caroline! this is good, take it down.' "[102] Caroline had bagged her game. She wrote up her material over the next year, and the book was published in the fall of 1934.

Not all her projects from this period of 1933 and 1934 came to fruition, however. She tried to do a series of articles for a Sunday School publication. In addition, based on an incident in Murfreesboro, Tennessee, she planned a novel about two maiden sisters living together; one comes slowly to realize that the other has been murdering their relations for the insurance money.[103] Caroline was also working on "an expatriate story from what I suppose is a Southern slant, niggers, or rather one nigger in Paris."[104] Neither of the Tates was ahead of their times in their racial views. Typical of their age and class, they regarded blacks with a kind of paternalistic, amused condescension, but, with their country, their attitudes became more liberal over the years.

Other projects from this time were more successful. "Old Red" was published on both sides of the Atlantic, in September 1933 in

T. S. Eliot's English *Criterion* and in December in *Scribner's*. The story was reprinted in the *O.Henry Prize Stories of 1934* where it received the second prize. *The Magazine* took the possum hunt chapter of *Aleck Maury, Sportsman* for its issue of April 1934 where it appeared as "What Music." Although she could not finish *The Garden of Adonis,* in the autumn of 1933 she turned one of its chapters into a short story, "A Morning's Favour," which was published by the *Southern Review* in 1935.

Allen, too, was having a period of mixed results. He was not writing much poetry, and would not, until the next decade. He did, however, win the Midland Author's Prize in October 1933, but his joy was dimmed by his father's death and its manner: John Tate was struck by a car in Cincinnati on October 21. Allen attended his funeral in Cincinnati, unaccompanied by Caroline. That fall Allen finally admitted that he could not bring himself to finish his biography of Robert E. Lee and added the advance from the publisher to his debts. Caroline approved; "It has got to the point where neither of us could stand it any longer, the nervous strain, aside from the unpleasantness of slowly starving to death while Allen tried to write a book he never wanted to write."[105]

The Tates did not face the prospect of "starvation" alone. As usual, they had a full complement of visitors. Caroline's Aunt Piedie, who had taken Nancy while the Tates were in France, made her annual autumn visit to her kin. Caroline's other maternal aunt, Loulie Meriwether, decided to move in with the Tates and take over the housekeeping while Caroline finished her novel. The house required plenty of keeping over the Christmas holidays since the Tates filled it with visitors, including Loulie's daughter Manny, the Warrens, the Ransoms, and others. They played dice, charades, and poker. All the gaiety may have been too much for elderly Aunt Loulie since she passed out after drinking two glasses of eggnog. After the holidays, Caroline experienced her usual January depression: the cook left, Allen had the flu, and Caroline had to milk the cows.[106]

The New Year woes lasted into the spring. On April 1, 1934, Caroline and Allen's car was struck by a car which was coming out of a side road. Their car turned over repeatedly, but, almost miraculously, neither was seriously injured. The knob for the ignition jabbed into Allen's leg and Caroline suffered from a wrenched shoulder for the next month. Both were extremely glad Nancy was not with them at the time. Although the Tates were relatively unscathed, their car was beyond repair so they were compelled to add the cost of a replacement to their mounting debts.[107]

As in 1932, the Tates were unsuccessfully attempting to maintain a large house and entertain openhandedly on the uncertain income of freelance writers. Again, they knew they should leave Benfolly until their finances improved. Although no foundation came to their assistance this time, a friend did. Robert Penn Warren was leaving Southwestern University in Memphis for Louisiana State University. Allen was offered Warren's old position, and the Tates, including Nancy, moved to Memphis in the fall of 1934.

Although they obtained what Caroline called a "good Confederate address," 2374 Forrest Avenue, they felt the bungalow was something of a step down after Benfolly. The house had only one floor which contained a living room, dining room, kitchen, small bath, and two bedrooms. Despite its comparatively compact dimensions, Caroline opted for household help. On October 1, 1934, she wrote, "I just got a cook this morning. I tried doing the work myself. I could manage all right and have plenty of time left over to write but as Hart [Crane] said, 'it constricts my imagination.' When I get through the ideas that seemed so hot in the morning aren't worth fooling with."[108] Without her rural occupations, Caroline also found it "just dull enough here to afford sufficient diversion and yet not distract you from your work."[109]

That fall she managed to complete two more Aleck Maury stories, "To Thy Chamber Window, Sweet" and "The Last Day in the Field." Both would appear in *Scribner's,* in December 1934 and March 1935, respectively. In "To Thy Chamber Window, Sweet," Aleck Maury narrowly escapes the toils of matrimony in the form of an attractive widow at his boarding house. He associates her admonitions about his clothes and diet with his discovery that in the local lake any fish he hooked was " 'snarled up in eel grass so tight he can't budge.' " Maury decides to evade both snares and slips away at night.[110]

In "The Last Day in the Field," Maury confronts a snare which he cannot escape, his own mortality. As he rests his bad leg, before returning home, he decides to take a last shot at a bird. The quiet elegance of Gordon's autumnal imagery conveys a sense of how the autumn of life feels to Aleck Maury. "I saw it there for a second, its wings black against the gold light, before, wings still spread, it came whirling down, like an autumn leaf, like the leaves that were everywhere about us, all over the ground."[111]

For Caroline, the main event of the fall of 1934 was the publication of *Aleck Maury, Sportsman*. She was displeased with the reviews since she believed they were often too short and written by people who had either not read or not understood the novel.[112] The sales, too, she found disappointing: only 1,300 copies by February 1935.[113] Despite the failure of her expectations, she almost decided to write another similar book. In the nearby Hickory Valley lived a seventy-year-old man who had devoted his life to sport, including bear hunting.[114] Although this project came to nothing, she did write another sportsman's story, with the assistance of another local sportsman, Nash Buckingham. "B from Bull's Foot" was published in the August 1936 issue of *Scribner's*.

Caroline did start another novel during that first academic year at Memphis. By June 1935 she had written three chapters of "The Cup of Fury," which was not another sportsman's sketch but the beginning of a panoramic account of the Civil War. Although she wrote that she intended "to barge right through, paying very little attention to the writing of it," the novel did not appear until 1937 under the title *None Shall Look Back*.[115]

Caroline was not all work and no play, however; as usual, the Tates had an active social life. On Sunday nights they hosted the faculty playreading club.[116] Caroline also joined the faculty walking club although "Allen won't walk, of course, but attends the tea-supper that follows the walk."[117] Caroline is alluding to what she considered Allen's physical laziness; apparently walking was ranked with gardening in Allen's list of strenuously boring activities. He was not equally averse to driving to Nashville, so the Tates made weekend trips there and to Merimont and Benfolly. "We are driving up to Benfolly this week end to see about the corn crop and the dogs and cats," Caroline wrote to Mark Van Doren in November 1934. "I am buying myself a winter coat, and a good one too, out of the corn crop. Let no one say now that the Tate family isn't consistently agrarian!"[118]

In spite of Caroline's agrarian bravado, the Tates also needed some urban pleasures. They planned a trip to New Orleans and

Baton Rouge over the Christmas vacation where they would see
their friends the Manson Radfords and the Robert Penn Warrens.
Apparently, the trip was a nonstop celebration, befitting south
Louisiana's party-loving traditions. On her return, Caroline wrote
to the Warrens, "Well we will have to pay two days for each one of
the revel but even at that it was worth it. There are no regrets. But
we are still marvelling at our endurance. Transferring the house
party from New Orleans to Baton Rouge and from gin to whiskey
without losing a member was really a feat that should go down in
the annals."[119]

The Tates felt this need to cut loose over Christmas because they
were feeling increasingly confined and burdened in Memphis. Allen was realizing how hard it is to combine full-time teaching with
serious writing. That fall of 1934 he was frustrated because he was
not making progress on his book of essays. Caroline seemed to feel
caged and repressed, as her references to some large cats in the
Memphis zoo seem to indicate. In October 1934, Caroline had
written, "It is so sad to see the black panther, though, and watch
his eye light as he gazes past you and sees a bird on a bough, or a
plump little boy." In February, she reported, "The black panther
at the zoo is dead, got in a fight with the puma . . . that looked so
sick." Caroline turned her attention to the lion: "As he got older
he lost his taste for ice cream and got savage so had to stop his
visits to the pavillion."[120] Throughout her life, Caroline was passionately fond of animals and would project her own feelings on
them. She would tell people what her dogs were "saying," which
would be a comic version of her own views.

In the spring of 1935, the Tates were exploring a possible way
out of the Memphis cage. Allen was negotiating with his wealthy
brother Ben to fund a new little magazine in Cincinnati that Allen
would edit. Both Allen and Caroline were ambivalent about the
necessary move to Cincinnati and close proximity to the Tate clan.
Caroline wrote, "Allen dreads the idea more than I do though I
can see how it would be a good deal of Tate to have around."[121]
The project never materialized, however, so the Tates had to seek
other ways out of their Memphis career.

Inadvertently, they were almost forced out that spring. James
Rorty, a northern journalist, asked Allen to take him to Marked
Tree, Arkansas, the scene of a labor dispute. Caroline did not

want Allen to go, but when he decided he would, she decided to accompany him. Rorty apparently managed to antagonize the town fathers with his prosecutorial style of questioning. On their return to Memphis, Allen was called into the presence of Southwestern's president, where he also found Marked Tree's preacher, plantation boss, and mayor, as well as the filling station operator who had heard some of the questioning. Caroline related the outcome in a letter to Sally Wood.

> The preacher and his cohorts threaten that if any story is written they will give a story to the A.P. saying as how this northern agitator accompanied by a Southwestern professor stopped him on the street (blocked his passage on the street) and questioned him against his will. . . . The college was just on the verge of a drive for funds to carry on with. There isn't anything Allen can say of course except that he doesn't think James will write a story.

Caroline found Allen's position even more absurd since "neither Allen nor I believe that northern agitators ever do anything but harm in the south."[122]

The Tates made a happier excursion that spring when they attended the "Conference on Literature and Reading in the South and Southwest" in Baton Rouge on April 10 and 11, 1935. The conference was held to celebrate Louisiana State University's seventy-fifth anniversary and inaugurate the *Southern Review* under the editorship of Robert Penn Warren and Cleanth Brooks, then professors of English at LSU, and the graduate dean, Thomas Pipkin. Although the conference title was broad, much of the discussion concerned the direction the *Southern Review* should take. The Agrarians present and their allies wanted it to be the nucleus of a southern publishing center that would make southern writers independent of New York, while others, such as the New Orleans writer Lyle Saxon, felt that "New York has treated us perhaps better than we deserve."

Both the Tates were invited to the Conference as was their old friend Ford Madox Ford. Allen led a discussion on what the individual Southern writer should do, considering the lack of a reading public in the South. Caroline, in contrast, was not in the spotlight. Her only recorded remarks occurred during the Third Session on Thursday. In the Second Session, Ford had argued that "The job of the editor has to be to get out good work. . . . If you want to have a southern magazine, let it not be necessarily southern but simply good." In the Third Session, Ford continued his defense of quality regardless of place, saying, "You must have a social center where you can meet your fellows in your craft. New York is such a place." Caroline disagreed, stating, "The southern writer has no chance in New York for such contact with fellows of his craft," seemingly forgetting her own New York contact with Katherine Anne Porter and Josephine Herbst in her desire for a Southern literary center.

In hindsight, Ford's direction for the *Southern Review,* "not necessarily southern but simply good," seems to have prevailed as did, eventually, his criticism about women's lack of role in the Southern literary establishment. "I am surprised that you had no woman address this meeting like Caroline Gordon and Elizabeth Madox Roberts."[123]

She did not seem to mind her lack of prominence and wrote that she "had a grand time." Perhaps Caroline was simply content to have been included as a participant in Baton Rouge after her rankling exclusion from the Southern Writers' Conference in Charlottesville in 1931. After Allen took the train back to Memphis, Caroline continued her holiday with Andrew Lytle's sister Polly. They drove to New Orleans and spent almost a week "a-pleasuring" themselves.[124]

The *Southern Review* conference was not the only alleviation of the Tates' routine that spring. As Caroline announced to Sally Wood in March, Allen "has been able to do a little work lately which causes great rejoicing in the household." In a letter two weeks later, she returned to the subject, "Allen is at last working again. If a paralytic had suddenly thrown away his crutch to take dance steps we couldn't be more pleased. It has considerably lightened the thick murk of gloom which has hung over this household for two years now."[125] Two years ago Allen had an affair with Marion Henry. In this letter, Caroline is making the link she usually did between Allen's dry spells as a writer and his infidelity, although which is cause and which effect is never quite clear. In any event, Allen had written most of his *Reactionary Essays on Poetry and Ideas* (1936) by the end of the first academic year in Memphis.

Aside from his heavy workload, Allen's scorn of local literary women temporarily cut him off from some sources of inspiration. Danforth Ross, Caroline's Clarksville cousin, taught at Southwestern after the Tates left; he found Allen's attitude was still remembered. Allen "was asked to address a group of lady poets. During the course of his meeting with them, they asked him to read and criticize their poetry. He took one look at one poem. 'You ladies,' he said, 'have no business writing poetry. You should be at home darning sox and knitting and cooking meals, doing things you understand.' "[126] Because of his horror of women amateurs, he "fought knowing" Anne Goodwin Winslow. When he finally agreed to meet her, he found her "a cultivated lady of the southern school and a beautiful writer of published prose." Caroline also liked her, calling her "one of the nicest people we've met here."[127]

One evening at Mrs. Winslow's house, Allen received the inspiration for his next work as he read aloud Henry James's classic ghost story, "The Turn of the Screw." His blocked imagination was once again engaged, and he collaborated with Mrs. Winslow in adapting the nouvelle into a play called *The Governess*. Despite the praise of drama critic Stark Young and others, the play was not produced at that time because, according to Allen, the James estate did not care for his interpretation and so would not grant an option to producer Herbert Shumlin. *The Governess* would not be produced until 1962 at the University of Minnesota.[128]

Caroline, too, found inspiration in an unexpected source. She consented to help a Memphis woman, a Mrs. Myrick, with a novel she was writing. Since Caroline regarded the project as a time-consuming burden, she was unenthusiastic at the time. She had, however, begun the task that would increasingly occupy the rest of her life: the teaching of creative writing, in and out of the classroom.[129]

Caroline was also planning to teach writing within the classroom since she and Allen were offered summer jobs at Olivet College in Michigan. Before their trip North, they returned to Benfolly and acted as if they planned to take up permanent residence. They painted, cleaned, and rearranged for several weeks. Company soon arrived to admire their efforts: Ford Madox Ford and Janice Biala, and some friends from Memphis, including Theodore Johnson, one of Allen's Southwestern colleagues. After the influx, the

Tates closed Benfolly and set off for Olivet by way of Louisville, where they learned from a telegram that their classes were cancelled for lack of pupils.

Curiously enough, the Tates did not return to the newly refurbished Benfolly. Caroline maintained that it would be more trouble than it was worth to reopen the house and get a maid before school began in the fall. Instead, by the end of July, they had taken refuge at Cornsilk, the farm belonging to Andrew Lytle's father in Guntersville, Alabama. The move was a successful one for Caroline. She was relieved of housekeeping responsibilities by Andrew's active father and his factotum. Caroline wrote in astonishment, "I never realized how much work you could get done visiting. . . . I feel rather guilty not being at Benfolly and visiting instead of being visited but it's grand. I like it."[130]

In addition, Caroline had the boon of her favorite writing companion, Andrew Lytle, who was also at work on a Civil War novel, *The Long Night* (1936). Caroline and Andrew established what they referred to as "THE routine" consisting of "breakfast, work, lunch, short nap, work, swim from five to six thirty, cocktail (one), supper, bed. Sleep like hell. Get up and go at it again."[131] Allen, however, did not find "THE routine" as beneficial because his energies were divided between a number of prospective projects. He took a respite from the rigors of Cornsilk to visit the Herbert Agars in Louisville, Kentucky. Andrew and Caroline pressed on, and Caroline made great progress on "The Cup of Fury" before the Tates returned to Memphis in the fall.

Caroline was able to maintain her pace in Memphis, perhaps because of the momentum she had built at Cornsilk or perhaps because she now had a congenial working environment. The Tates had moved to another bungalow at the other, "leafy" end of Forrest Avenue. Caroline wrote, "Allen calls it our gold oaken nest and it is pretty full of g.o. But there is a heavenly work room glassed-in sun parlor at the back of the house which we both enjoy." Indeed, Caroline was so absorbed in the Civil War that she seemed to have lost some perspective on the present. When she responded to Nancy and Allen's complaints about their dinner with an account of Confederate soldiers subsisting on blackberries, Nancy replied, "Mama, I don't care what those old Confederate soldiers ate."[132]

Work was interrupted for the Christmas holidays at Benfolly, where the Tates and the Meriwether kin were joined by Red and Cinina Warren. After the holidays, the Tates returned to Memphis. In February 1936 Caroline's neglected novel received new impetus from the visit of Andrew Lytle and his sister Polly. Caroline wrote to Sally Wood that the "house is turned into a regular book factory. . . . Andrew is working on Shiloh, I on Chickamauga. We compare symptoms and it helps a lot." Allen also participated in this work party; according to Caroline, he was experiencing "a renascence" and "turning out essays like a house a fire."[133] His essays would appear that year as *Reactionary Essays on Poetry and Ideas*. The year 1936 was indeed one of accomplishment for him since he also published *The Mediterranean and Other Poems* and participated in the new Agrarian symposium, *Who Owns America?*

Although Caroline did not finish her novel in time for 1936 publication, she did publish a short story, "One More Time," in the December 1935 *Scribner's*. Once again, she uses the voice of Aleck Maury, this time to tell the story of an incurably ill fishing companion who drowns himself rather than die a lingering death without the pleasure of sport. Maury's confidence in the future in which "tomorrow was *bound* to be a good day" is undercut by the appearance of his terminally ill friend. Without realizing its significance, he repeats his friend's advice about willow flies: "Here today, you know, and gone tomorrow." The reader knows that the words are true for the aged Aleck as well, and that, with his unquenchable zest for sport, he too will long for "one more time."[134]

During that spring of 1936, Caroline took time out from her novel to write what she called a "horror" story. "The Enemy" is her only published horror story and her only tale exclusively concerning blacks. The parents and husband of a murdered woman wait for news of the murderer's execution. When the husband learns that the murderer showed no penitence, he kills himself and his hunting dog to avenge his wife's honor in the next world.

Unlike Faulkner, Caroline did not attempt to imagine the black experience from inside. Perhaps realizing her own limitations, she chose an omniscient narrator, making "The Enemy" the only story in her first collection, *The Forest of the South* (1945), that is not told by a first-person narrator or through a central consciousness.

The narrative voice is the chorus-like one of traditional ballads, which are also tales of rural love, violence, and grief. Her characters contribute to the effect with repeated utterances that resemble refrains, as critic John E. Alvis has pointed out.[135] The choice of black characters may have more to do with her desire to achieve this rural balladic effect than any particular desire to address questions of race.

Caroline further detached herself from the story by signing Allen's name to it before she sent it to *Esquire*. Although her action may have been an attempt to appeal to *Esquire*'s masculine bent, it may also disclose her ambivalence about the tale. In any event, her ploy did not work since the story did not appear in *Esquire,* but under her own name in the Spring 1938 issue of the *Southern Review.*

After the Lytles left, the Tates had another important visitor, their old friend from Greenwich Village, Dorothy Day. Caroline gave Dorothy Day's biographer this account of her arrival in Memphis. William D. Miller writes:

> What Dorothy wanted was a bath, a clean bed, and some catching up on the news of what was going on in the life of their mutual acquaintances. . . . Caroline, with mock severity—although she meant it, too—told her that, because Dorothy was a communist and because she had experienced enough South-baiting from Dorothy's leftist friends, that Dorothy could not come. To which comment Dorothy "laughed heartily." She told Caroline that she had become a Catholic.[136]

Not only had she become a Catholic, but she was a founder of the Catholic Worker movement and the editor of its newspaper, also called *The Catholic Worker.*

Dorothy Day was particularly eager to see the Tates because she had recently discovered *I'll Take My Stand* and recognized the similar aims of the Agrarians and Catholic Workers. In her autobiography, she explains the Catholic Worker philosophy as expounded by her co-founder Peter Maurin:

> He wanted them [men] to stretch out their arms to their brothers, because he knew that the surest way to find God, to find the good, was through one's brothers. Peter wanted this striving to result in a better physical life in which all men would be able to fulfill themselves, develop their capacities for love and worship, expressed in all the arts. He

wanted them to be able to produce what was needed in the way of homes, food, clothing, so that there was enough of these necessities for everyone. A synthesis of "cult, culture and cultivation," he called it.[137]

With the exception of the Catholic "cult," these were the aims of the Agrarians.

The Tates each responded to Dorothy Day as they would to their conversions to Catholicism over a decade later. Allen was impressed by her theoretical grasp of the similarities betwen Agrarianism and the Catholic Worker movement as well as her zealous devotion to her ideals: "A very remarkable woman. Terrific energy, much practical sense, and a fanatical devotion to the cause of the land."[138] Characteristically, Caroline seemed more interested in how Catholicism affected her character and her way of life. Nancy Tate Wood remembered that her mother was attracted to Dorothy Day because she was so calm and peaceful. Caroline was also genuinely interested in the Catholic Worker movement and contributed money to it.[139]

Neither Caroline nor Allen could be "calm and peaceful" yet since, as usual, they needed to earn some money. Benfolly remained closed as the Tates spent the summer of 1936 on the move. First Allen delivered the Phi Beta Kappa address at the University of Virginia. Next, they drove to Michigan where they participated in the writers' conference at Olivet College for two weeks. According to Caroline, "We were—and the conference was—a great success." Through the good offices of Mark Van Doren, critic, poet, and Columbia English professor, Allen obtained a lectureship at Columbia University where the young poet John Berryman was one of his students. Caroline wrote that in New York they "ate too much, drank too much" and saw old friends such as Max Perkins.[140]

When Allen accepted the teaching position at Southwestern two years before, the Tates believed that teaching was the only alternative to life at Benfolly. They decided to try a third way of life for the academic year 1937–1938 by moving in with Andrew Lytle at his cabin at Monteagle, Tennessee. The Monteagle Assembly was a Chatauqua-like community, mainly of summer houses, whose denizens participated in a round of lectures, religious services, and activities. In contrast, the winter was solitary and quiet. By pooling their resources with Andrew, the Tates could devote themselves to writing while living comfortably yet frugally. The menage was assisted by Andrew's father who regularly brought food, cooks, and assistance from his farm.

"We will probably never open Benfolly again till one of us writes a best seller," Caroline wrote to Sally Wood in September.[141] Although Caroline had been working on her Civil War novel, *The Cup of Fury,* for several years, she was probably referring to the phenomenal success of Margaret Mitchell's *Gone With the Wind* (1936). Allen was also at work on a Civil War novel, *The Fathers* (1938). Caroline, in particular, was worried that her novel would be lost in *Gone With the Wind*'s shadow. "Margaret Mitchell has got all the trade, damn her. They say it took her ten years to write that novel. Why couldn't it have taken her twelve?"[142] Sales competition aside, neither of the Tates thought much of *Gone With the Wind* as a novel. Nancy Tate Wood remembered her father repeatedly saying that *Gone With the Wind* had set back the technique of the novel by years.[143]

Despite these worries, Caroline forged ahead and was reading final proofs of her novel by the end of November in expectation of publication early in the New Year. She wrote to Robert Penn Warren, "We have had a movie nibble—nibble not strike—nothing to it probably but it is easy enough to turn our imaginings towards the coast—but it's easy to touch off our imaginings as you know." The lure of Hollywood remained unfulfilled, and the movies became a source of annoyance. Caroline wrote to Warren, "it is not to be 'The Cup of Fury.' Metro Goldwyn Mayer owns said title and if I used it they could get out a picture under that name without paying me a cent. Hence, much tearing of hair, thumbing of the Bible, Shakespeare and Milton. Can't get a title that's any good. I reckon they'll call it 'None Shall Look Back.' From Nahum if you've heard of such a prophet."[144] Warren was kinder than Hollywood and included a chapter from the novel, "The Women on the Battlefield," in the winter issue of *The Southern Review.*

Caroline did not spend the entire fall of 1937 worrying over and correcting *None Shall Look Back.* She wrote a fond and appreciative sketch of Andrew Lytle's father, which remained unpublished after its rejection by the *Virginia Quarterly Review* because it was not their kind of material.[145] She also completed a short story set at Monteagle, "The Brilliant Leaves." A tale of a tragic lack of understanding between lovers in an autumnal forest, "The Brilliant Leaves" displays Caroline's natural symbolism at its best. Although men refuse to accept the message of the brilliant leaves, women recognize that all things must mature and die. As in "Old Red," for women, time is essentially static in that everything fits into a larger pattern that is unchanging.

"The Brilliant Leaves" opens with Jimmy's mother and aunt weaving the lives of the people in the white houses into a pattern with their talk. They turn to years past when "high spirited" Sally Mainwaring had "climbed down a rope ladder to meet her sweetheart while her father stood at another window," shotgun in hand, but when "she got to the ground the lover had scuttled off into the bushes."[146] This anecdote sets the pattern for the action that follows. Evelyn wants to follow the progress of nature to a mature love, but Jimmy is unable to rise to the challenge. Like Aleck Maury with his sport, Jimmy wants to seize and hold the love of June and refuses to recognize that his adversary, Time, will win.

Nature mirrors the lover's reality.[147] Evelyn and Jimmy meet in the autumnal forest where their love had grown in June. When Evelyn comments that "It's different," Jimmy refuses to acknowledge the truth of her remark. He replies, "I know a place where it's still green. . . . I was there the other day. There's some yellow leaves but it's mostly still green. Like summer." In that green spot, Jimmy makes a sexual advance, but Evelyn wants something more. She wants to try to achieve something together as mature lovers would, in this instance to climb beyond the significantly named "Bridal Veil Falls." When Evelyn loses her footing and falls, Jimmy is unable to save her. As is often the case in Caroline's fiction, man's absorption in his own needs and desires leaves woman vulnerable to a tragedy from which he cannot save her.

With imagery that reinforces its major themes, Caroline ends "The Brilliant Leaves" with Jimmy running away for help as Evelyn lies alone and dying among the ferns.

> He ran and the brilliant, the wine-colored leaves crackled and broke under his feet. His mouth, a taut square, drew in, released whining breaths. His starting eyes fixed the ground, but he did not see the leaves that he ran over. He saw only the white houses that no matter how fast he ran kept always just ahead of him. If he did not hurry they would

slide off the hill, slide off and leave him running forever through these woods, over these dead leaves.

Jimmy persists in his masculine refusal to heed the message of the dead leaves and continues to try to outrun time. Significantly, he is running toward the white houses that contain the women who can place his tragedy in the stable pattern of events. Without them, he will be "running forever, through these woods, over these dead leaves." With powerful imagery, Caroline once again represented the hopelessly incongruent aspirations of men and women that are the key to her work.

"The Brilliant Leaves" was rejected by the *Virginia Quarterly Review* because its editor felt that Evelyn's death seemed gratuitous and the reader was not made ready for it. Robert Penn Warren at *the Southern Review* liked the ending, but also did not believe that it integrated what had come before.[148] Both these editors were correct; read outside of the context of Caroline's works as a whole, the forceful imagery is much more difficult to decipher. The story was eventually accepted by *Mademoiselle* where it appeared in July 1937.

Caroline was distracted from her concern over this story and the upcoming publication of *None Shall Look Back* by the Tates' travels over the Christmas holidays. They visited Nancy in Chattanooga where she was spending another year with the Campbells, the family of Caroline's Aunt Piedie. After a brief visit to Merimont, the Tates headed for Richmond where Allen addressed the annual meeting of the Modern Language Association. Her encounter with Thomas Wolfe seems to have been Caroline's most important memory of the occasion. "He was drunk and extremely amiable. He kept looking at me and blubbering 'Mrs. Gordon, Max Perkins thinks you're wonderful.' He is so dumb that he can hardly follow a conversation. We were talking about the wonderful whorehouse scene in [Faulkner's novel] Sanctuary. Wolfe assured us that he had intimate acquaintances with whore houses in many places and that whore house wasn't true to life."[149]

After a brief trip to Washington, the Tates returned to Monteagle to continue their work and await the critical reception to *None Shall Look Back* which was to be published in February 1937. The core of *None Shall Look Back* is based on the antithetical character of Caroline's maternal grandparents, Caroline and Douglas Meriwether, who appear in the novel as Lucy and Rives Allard. The recently wed Lucy loves Rives, but he has dedicated himself to the god of war in the guise of the Confederacy. He progressively divests himself of any human ties and pleasures and even accepts the solitary and dangerous mission of a spy although it prevents any communication with his wife. On his last leave at home, Rives shouts in his sleep, " 'stick him and leave him.' "[150] Lucy tries to understand the murderous stranger he has become:

> She felt a sudden revulsion from the man at her side. Raising herself on one elbow she had studied his face. The light coming in at the window illuminated his features: the high, aquiline nose, the eyes set in their deep hollows, the stern mouth. In the moonlight they were like marble. The kind of face that might be carved on a tomb (357).

Lucy not only foresees Rives's death, but realizes that he is already dead to all domestic ties and tranquil human pleasures.

The moving story of Lucy and Rives reiterates Gordon's central theme of the sexes' unresolvably conflicting points of view, but *None Shall Look Back* goes beyond this private theme to explore heroism in the public arena of war. Caroline had Tolstoy's *War and Peace* in mind as she alternated domestic and battle scenes. She wrote to Sally Wood: "each battle, it seemed to me, had to be treated in a different way or you'd get monotony. . . . I treated Fort Donelson in Plutarchan style, reserving some impressionism for Chickamauga."[151] Some of these public scenes are enormously effective, such as the two small boys watching the beginning of Chickamauga or General Nathan Bedford Forrest confronting his pusillanimous fellow officers at Fort Donelson. Their inclusion, however, is at the expense of the novel as an integrated whole.

The fundamental trouble with the public scenes of *None Shall Look Back* is exactly the problem Caroline exhibited in writing *Penhally*. In her ambitious attempt to master a variety of fictional techniques, she sometimes forgot the reader's need to follow and understand the action. Without an expert knowledge of the Civil War, the reader cannot know what the battle is, why it is important, or even the date. In a way, this lack of information contributes to a sense of confusion that mimics that of war; a soldier, however, would at least know where he was, the name of his general, and some of his larger purposes.

One of the most fascinating aspects of Caroline's early fiction, however, is her ambition, the way one can see her strive to master techniques from work to work. She is not content to retell *Penhally*'s family saga in a more effective manner through the Allards, but adds further challenges. The locally focused, highly ritualized genre of modern Civil War fiction is too narrow for her; she wants to write an American *War and Peace* filled with historical giants and panoramic set-pieces. In the past, of course, such works were much more characteristic of male novelists than their sister artists. Although each of Caroline's experiments is not entirely successful in itself, it always bears fruit in a succeeding effort, in this case her next historical fiction, *Green Centuries* (1941). For this reason, she is often, and justly, called a novelist's novelist.

Although Caroline did not have a best seller of *Gone With the Wind*'s dimensions, *None Shall Look Back* did very well, selling ten thousand copies by May, "enough to live on—frugally—for a year," according to Caroline.[152] The novel received widespread attention, and Caroline was gratified by the perspicacious admiration of friends and fellow writers. Stark Young wrote to her praising *None Shall Look Back*'s "nobility of tone," though he did mention the difficulties in establishing a pattern of action, the lack of which makes the novel hard to follow.[153] In a review for the *New Republic,* Katherine Anne Porter recognized Gordon's inability to repeat herself and play it safe: "She might have done the neat conventional thing, and told her story through the adventures of her unlucky pair of young lovers, Lucy Churchill and her cousin Rives Allard. But they take their place in the midst of a tragedy of which their own tragedy is only a part." Porter also noted the advances Caroline had made in her art, writing, "This seems to me in a great many ways a better book than 'Penhally' or 'Aleck Maury, Sportsman.' "[154]

Some of the most interesting comments were made by Ford Madox Ford in some pages that survive in Caroline's folder at Scribner's, probably a pre-publication publicity statement. He writes of "her calm self—the calm self that gets into her writings. Her outside-the-study activities are on the side of what the French call *bruyant*—

a portmanteau word implying at once brilliant, clamorous, with a slight soupcon of exaggeration in all three attributes." He contrasts the carnage of the battlefield scenes with Caroline's authorial stance: "And beside you, Mrs. Tate remains mysterious, unimpassioned, almost impartial as the tragic destiny unrolls itself beneath you both. . . . It is as if she were Pallas Athene, suspended above the Greek hosts, knowing what destiny decrees." Ford returns to this theme a third time when he commends Caroline for knowing "that if your approach to horror is not that of the quiet and collected observer and renderer you will fail in attaining the real height of a tragedy."[155]

Ford is protesting too much as he attempts to praise *None Shall Look Back*. As he does so, he delineates one of the characteristics of Caroline's fiction that make it difficult for her readers to become involved in her work: her apparent lack of sympathy or feeling for her characters can elicit the same response in her readers. Many who knew her personally have commented that they always found her letters more engaging than her fiction because "the *bruyant*" side of her personality remains in the letters but is absent in her works. Also strikingly absent from her fiction is the sense of humor, of life as comedy, that appears in her letters and the reminiscences of those who knew her. The novels appear to be written by a woman with a tragic sense of life combined with a steely Olympian detachment. This split is indicative of the conflict in Caroline's character between what she perceived as "feminine" and "masculine" roles: women please by making life's woes into comic anecdotes and men instruct through presenting the tragedy unflinchingly.

Max Perkins, a staunch admirer of Caroline's work, was delighted with the success of *None Shall Look Back*. All the responses he received were not commendatory, however. G. Forrest Gillet of the Chicago Paper Company wrote to Scribner's pointing out two passages in *None Shall Look Back* that are very close to two passages in Andrew Lytle's *Bedford Forrest and His Critter Company*. More seriously, Scribner's also heard from J. W. Hiltman, Chairman of the Board of D. Appleton-Century Company, publishers of *Battles and Leaders of the Civil War*. He enclosed six pages with parallel columns from *Battles and Leaders* and *None Shall Look Back,* such as:

NSLB, p. 262: The Rebels in massed lines were already swarming around his flank. . . . Granger saw the Rebels strike his last brigade as it was leaving the line. It slammed back like a door and was shattered.

B&L, III, 663: The massed lines of the enemy swarmed around their flanks. . . . they struck his last brigade as it was leaving the line. It slammed back like a door and was shattered.

Scribner's replied that Gordon had taken factual matters from *Battles and Leaders* and that her source should have been acknowledged and permission asked, had they known about it. No further correspondence about the issue remains in Caroline's author file at Scribner's.[156]

Caroline may well have been ignorant of the need for correct citation in fiction, but as these representative quotations indicate, she seems to have used some descriptions as well as facts. She tended to rely too heavily on her research and factual materials because she lacked confidence in her imagination and found the research much more pleasant than the actual writing. She may well have learned from this experience since her next historical fiction, *Green Centuries* (1941) did contain an acknowledgement of her sources at the end.

No research was required, however, for the book Perkins was urging her to finish for fall publication. *The Garden of Adonis* is set in the environs of contemporary Clarksville. As Caroline had decided when she planned and began it in 1931, the novel initially concerns the irreconcilable conflicts of the planter and his poor white tenants. At the end of *The Garden of Adonis,* the planter Ben Allard is killed by his tenant Ote Mortimer in a dispute over when to mow some hay. Although their conflict is agrarian, their motives actually arise from various instances of adultery and infidelity. Ote Mortimer has just learned that his fiancée, carrying his unborn child, has eloped with the prosperous bootlegger Buck Chester. At the same time, Ben Allard has discovered that his unmarried daughter Letty has run off with the married Jim Carter. Jim Carter was responding to the adultery of his wife Sarah who, in turn, had been acting in retaliation for Jim's affairs with two local women. Jim's tangled and sordid history takes up the central third of the book, interrupting the Ben Allard–Ote Mortimer main plot.

The structural problems of *The Garden of Adonis* are obvious, as are the reasons behind them, when the history of the novel's composition is considered. Caroline abandoned the novel when she learned of Allen's adultery in the spring of 1933 and took refuge in her father's consciousness in *Aleck Maury, Sportsman.* To profit from the success of *None Shall Look Back,* at her publisher's urging, she returned to her abortive novel early in 1937. In the interim, the focus of the book had shifted from the tenant system to betrayal in half a dozen permutations. Unsurprisingly, Caroline considered *The Garden of Adonis* her least favorite novel. Structurally it is her weakest, though the novel does contain some stunning descriptions and characterizations.

This was the problematic work Caroline proposed to finish in the summer of 1937 as Allen was trying to finish *The Fathers.* Sensibly, the Tates decided to retreat to Benfolly, but less sensibly perhaps, they invited a houseful of guests. Allen and Caroline arrived at Benfolly at the end of April and indulged in their usual frenzy of cleaning and refurbishing in preparation for the Fords' arrival early in May.

The Tates' domestic labors were interrupted by the appearance of a young Bostonian, Robert Lowell. At the behest of his psychiatrist, the former Fugitive Merrill Moore, Lowell had come South to sit in on some of John Crowe Ransom's classes at Vanderbilt. If he enjoyed them, the budding poet was to follow Ransom to his new position at Kenyon College in Gambier, Ohio, since Lowell was unhappy at Harvard. Before heading for Nashville, Lowell arrived at Benfolly to introduce himself to the Tates, but his main objective was to see Ford Madox Ford, whom he had met and idolized that spring in Boston. Lowell recalled his appearance on the scene.:

> My head was full of Miltonic, vaguely piratical ambitions. My only anchor was a suitcase, heavy with bad poetry. I was brought to earth by my bumper mashing the Tates' frail agrarian mail box post. Getting out to disguise the damage, I turned my back on their peeling, pillared house. I had crashed the civilization of the South.[157]

Although the older Lowell's reminiscence is full of double-edged irony, a letter Caroline wrote at the time asserts that Lowell left a different mark on the civilization of the South:

> The other day we had what I believe is the strangest visitation we ever had. Allen and I were standing in the circle admiring the lemon lilies

when a car drove up to the gate and a young man got out. He stopped down there by the post box and answered the calls of Nature then ascended the slope. We stood there eyeing him sternly and were on the point of shouting "defense d'uriner" when he came up to Allen, regarded him fixedly and muttered something about Ford.[158]

Whatever impression he had made, the young poet was impressed by "the stately yet bohemian" Tates and wanted to return when Ford arrived.[159] The Tates knew that every bed would be taken and believed they had dismissed the importunate Lowell by telling him the house would be so full he would have to erect a tent on the lawn.

Caroline did not regard the Fords' advent with Lowell's unmitigated enthusiasm. On their last visit, there had been some fuss over Ford's professed inability to eat anything but French cooking. Caroline was determined that they would eat what she could provide: Tennessee cooking produced by the Tennessee hostess and her Tennessee servants. She fumed, "When I think of all the people I've had here, feeding them on a shoe string (financially if not literally) . . . but none of them ever really complained."[160] As it happened, Ford arrived with "insomnia, indigestion, and gout to boot," and Caroline was forced to provide a special diet.[161]

The tempers of the ménage were further exacerbated by political, artistic, and sanitary concerns. The Fords had with them Biala's sister-in-law, Mrs. Jack Tworkov, who acted as Ford's secretary. Both Biala and Mrs. Tworkov held political opinions much to the left of the Tates, which was not difficult considering the Tates' conservatism. Aside from the awkwardness of inharmonious views, Mrs. Tworkov apparently feared she or the Tates would be lynched if *The Daily Worker* were found in their mailbox. If Mrs. Tworkov felt so isolated by her politics, Biala felt isolated by her metier. A painter, she was surrounded by writers who were either clattering on their typewriters or discussing what they had just hammered out. She wrote, "It is awful here. In every room there's a typewriter and at every typewriter sits a genius. The genius is wilted and says that he or she can do no more but the typewritten sheets keep on mounting."[162]

The geniuses were "wilted" not only by their literary enterprise, but by the drought that was multiplying the effect of the summer heat. Since Benfolly was dependent on cisterns, the water supply was brief and uncertain. Allen became angry at the Ford group for flushing the toilets too often. Ford decided to vindicate himself by constructing a dew pond. As Lowell's biographer Ian Hamilton relates, "He sank a bathtub in a nearby meadow, filled it up with twigs, and was baffled and outraged when it failed to produce a drop of liquid."[163]

Ford's discomfiture was exacerbated by the presence of Robert Lowell as an onlooker. Lowell had returned from Nashville where he had taken the Tates' jocular speech to heart and purchased a tent which he proceeded to erect on the Benfolly lawn. Ford, probably jealous of the attention the young man was receiving from Allen and Caroline, was not happy to see him and refused to speak to him or acknowledge his presence for days. Carorline could stand only so much of this and gave Ford "hell about not speaking to him and he now addresses him as 'Young Man.' "[164]

As this anecdote shows, Caroline liked and pitied the young poet. One reason may have been that he assisted her in the work of feeding the bodies that pushed the typewriter keys and paintbrushes. "He's such a nice boy. Drives out to Merry Mont to haul in buttermilk, etc., flits the dining room—the handiest boy I ever knew in fact. When he isn't doing errands he retires to his tent whence a low bumble emerges. Robert reading Andrew Marvell aloud to get the scansion."[165] In fact, Lowell acquitted himself so well that he was invited to accompany the party to the Olivet Writers' Conference in July and serve as Ford's secretary.

At the Writers' Conference, the Tates met their old friend Katherine Anne Porter. They wanted time to continue their reunion so Katherine Anne returned to Benfolly with the Tates, stopping on the way to see some historic homes in Virginia.[166] At Benfolly, Katherine Anne took to rural living with a vengeance, as Caroline recalled:

> She couldn't write here—life was so distracting, what with the cats and all the fruits of the earth needing to be preserved, pickled or made into wine. She made mint liquor, preserved peaches whole, made five gallons of elderberry wine, brandied peaches and would have brandied and preserved bushes more if I had provided her with them. All this, of course, partly out of domestic passions, partly out of charitable concern for our welfare and a good part I wickedly believe just to get out of

work. She has to sever every earthly tie she has before she can do any
work, go off to a hotel somewhere usually.[167]

Caroline's description of Katherine Anne has a tinge of envy because Katherine Anne did not even attempt the sort of balancing
act between domesticity and art that Caroline was struggling to
maintain. After the amount of company she had entertained that
summer, Caroline may well have considered the ability to "sever
every earthly tie" desirable.

Katherine Anne, on her part, experienced more than "domestic
passions" on her visit to Benfolly. There she met Albert Erskine,
the young Business Manager of the *Southern Review*. Allen remembered them sitting on the verandah in the moonlight until the wee
hours, talking and talking, and keeping Allen up as their conversation drifted in his window. After her visit was over, Katherine followed Erskine to Louisiana where he became her next husband.[168]

When the Tates closed Benfolly at the end of the summer of
1937, they also closed an epoch in their lives. They would never
live there again nor would they ever again have as firm and treasured a home base. They were trading a settled home precariously
maintained by freelance writing for the financial security but roam-
ing lifestyle of academic appointments. Gypsy scholars they would
become and gypsy scholars they would remain until the end of
their lives.

Caroline Gordon at age ten in Clarksville, Tennessee. Her eyes look wet here because the photographer had just commented to her mother that it was a pity the boys, her brothers, were better looking. Caroline remembered this incident with chagrin in the last month of her life. (Courtesy of Nancy Tate Wood)

Caroline (far right) with young Meriwether kin. She looks unhappy as a member of the group, but is unable to detach herself completely, her characteristic, lifelong stance toward the Meriwethers. (Courtesy of Nancy Tate Wood)

Morris Meriwether Gordon, Caroline's envied and emulated older brother. (Courtesy of Nancy Tate Wood)

Nancy Meriwether Gordon, Caroline's cultured and enigmatic mother. (Courtesy of Nancy Tate Wood)

Allen Tate, age six, in November 1905. (Courtesy of Nancy Tate Wood)

Caroline (bottom right) again on the edge of a group, the freshman
class in the Bethany College yearbook of 1913. (Courtesy of Bethany
College)

Caroline graduates
from Bethany Col-
lege, from the 1916
yearbook. (Courtesy
of Bethany College)

"Allen and I, held together in space by Nancy," as Caroline prophetically described this 1929 portrait painted in Paris by Stella Bowen. (Courtesy of Nancy Tate Wood)

Caroline in Greensboro, North Carolina c. 1938. (Courtesy of Nancy Tate Wood)

Caroline and Allen in mid-life and mid-marriage with troubles multiplying. (Courtesy of Nancy Tate Wood)

Allen Tate in 1952 as photographed in Italy by Stephen Spender. (Courtesy of Nancy Tate Wood)

Caroline Gordon in
her early sixties.
Photo by Uli Stelzer.
(Courtesy of Nancy
Tate Wood)

Caroline as the aging
writer, vulnerable but
attempting a smile
and still trying to
write. Photo by Brad-
ford Bachrach. (Cour-
tesy of Nancy Tate
Wood)

CHAPTER 5

In the winter of 1938, Caroline Gordon began her career as a college teacher at the Women's College of the University of North Carolina at Greensboro. It was an auspicious start since her teaching load was light, only one class, and she could delight in teaching English and writing. She was no longer the sensitive novice of twenty years ago, struggling to teach uncongenial chemistry and French at the Clarksville High School. Now she could report to Maxwell Perkins that the "teaching racket is grand. The preparation consists mostly of lying in bed at night reading Tom Jones or something like that and the classes themselves take very little time."[1]

Not content with this easy "racket," she also advised budding creative writers at the Tates' home at 112 Arden Place. One such pupil, James Ross, recalls,

> My first impression of Caroline Gordon was that she was frank, direct, and went straight to the heart of the matter. And I sensed that she was generous; she'd take time and trouble to help other people. . . . As for her appearance, she was handsome in a sort of Celtic way, but sometimes she had a brooding air as if she mistrusted both the present and the future and was keeping her fingers crossed. Her manner was cordial . . . even when she was reprimanding me for some lapse of character or composition.

In particular, Ross remembers, she stressed that the novice fiction writer should study the techniques of Gustave Flaubert and Henry James to learn how "to cut out extraneous episodes or scenes, no matter how dramatic or entertaining, if their effect was to hold up the action." She told him, "In a short story, you should get things going in the first paragraph or set the scene in a way to nail down the reader's interest. And to keep going, you've got to be able to turn around on a dime."[2]

In some ways, Caroline Gordon's experience at Greensboro from January 1938 to June 1939 was typical of her next thirty-five years as a teacher. Without tenure, it was impermanent and brief. She expended a great deal of time and effort on extracurricular writing pupils, as she would do in the future. Her stress on technique over content became increasingly pronounced with the years as did her deification of Flaubert and James. She would, however, take on increasingly heavy classroom teaching loads so that her classroom preparation and performance became proportionately more important to her. The principal difference from her later teaching career can be attributed to her attitude. During her Greensboro years, she still considered her vocation to be writing. As her reputation became increasingly obscure over the years, she expended more and more of her energy on teaching as if she felt a need to confirm critical blindness toward her writing by enacting the old saw, "Those who can't do, teach."

This transformation, however, was more than a decade in the future, and at Greensboro, Caroline researched and began to write her fifth novel, *Green Centuries* (1941). She found inspiration in North Carolina and travelled around the state to discover the setting for her novel of pioneer life. She wrote to Maxwell Perkins, "It was a great stroke of luck to land in North Carolina just when I was writing this book—I've gotten a lot of stuff I never could have gotten in Tennessee. I am starting my settlers from Salisbury, not far from here, and have their route all figured out. Next month I am going over to the Watauga country and settle them in one of those valleys."[3]

Although she was exploring new settings in *Green Centuries,* Caroline was reworking some of her old themes with great power and effectiveness. She returned to the theme of women's abandonment by ambition-obsessed men through the character of Cassy, the wife of pioneer Rion Outlaw. As his name suggests he is a hunter, Orion, and feels a compulsion to travel beyond the confines of law and settlement. As Cassandra's name implies, she will be a victim of masculine battles. Caroline's original name for Cassy, Jocasta, also indicates that she, like the wife of Oedipus, must suffer and die as a consequence of her husband's hubris, in her case, Rion's overreaching ambition to keep travelling farther west in search of an even better place to settle.

A dark, proud, refined woman, Cassy suffers from the rigors of pioneer life. After the death of her children, she withdraws from her husband and turns to religion, a withdrawal similar to that of Caroline Gordon's own mother and that of Aleck Maury's wife Molly in *Aleck Maury, Sportsman*. When Rion is unfaithful to her, Cassy's sacrifices seem in vain. In a passage of painful intensity, which may derive from her own suffering at Allen Tate's affair with her cousin in 1933, Caroline explores Cassy's torments.

> Lying there beside him, thinking of that woman, she had gone cold as a stone and all the next day that moment kept coming back as if the woman walked behind her and now and then put out a hand and touched hers. She tried to drive it away by working but it kept coming back. And with the memory came the knowledge that Rion had endured the touch of that flesh, had sought it out, had gone to it again and again. "Like a dog to his vomit," she whispered to herself and standing at the churn her arms going up and down while her lashes beat off the tears she had whispered, "No, it was what he wanted . . . more than me . . . . She was what he wanted." And she saw herself as a creature of such poor account that a man would hardly go out of his way to avoid stepping on it and yet a creature that went on living its desperate life, like a rattlesnake she had seen once that some boys had caught and were teasing with Seneca root. The thing kept turning its head from side to side, loathing what was offered, and yet it had to go on, and every time it turned the boys would be there, holding the loathsome root up before it.[4]

Like Caroline's earlier heroines, Cassy turns "cold as a stone" in the face of her husband's infidelity, but keeps working with the stony-hearted precision of an automaton. What is new in Caroline's presentation of Cassy's response is her extreme self-loathing, characterizing herself as a "creature," "it," "a rattlesnake," and "the thing." Unlike the heroines of *Penhally* and *None Shall Look Back*, Cassy does not blame the impersonal forces of history, but feels herself at once guilty and powerless, the snake whose loathsome nature makes it unable to refuse the tortures of the equally loathsome boys in Caroline's Danteesque rendition of the vicious circle of heterosexual relations.

In *Green Centuries,* Caroline also returns to her theme of masculine restlessness or boredom in the face of the status quo. In *Penhally* and *None Shall Look Back,* the men turn to war for a challenge; in *Aleck Maury, Sportsman,* the title character obsessively pursues sport; and in *The Garden of Adonis,* the men fight over Agrarian ambitions and family honor as embodied in their women. In *Green Centuries,* Caroline explores that quintessentially American challenge to the status quo, the ever-beckoning frontier. Her analysis resembles that of Hector St. John de Crevecoeur, an eighteenth-century Frenchman, in his *Letters from an American Farmer.* Behind the persona of Farmer John, Crevecoeur posits the dangers as well as the promises of the frontier. If a person strays too far from settled society, he will revert to savagery; such a man is Rion's Outlaw brother Archy who is captured by the Indians and eventually joins the tribe. Unlike Crevecoeur, Caroline shows great knowledge of and sympathy toward, Indian customs. Her Indians are not "savages," but members of a settled society that is being destroyed by the greed and ambitions of whites, symbolized by Rion's slaying of Archy in battle.

As he pursues the American Dream, the alluring something better which is always just out of reach, Rion Outlaw is necessarily a man never at rest. Like Crevecoeur's frontiersman, he seems to acquire the worst of the two societies between which he moves, but can accept neither of them fully. As represented by his inability to read, the culture of civilized society is unavailable to him, and so holds no attractions. He is, however, touched sufficiently by it to make him unable to consider the Indian's freer life anything but inferior. In his inability to live in the past or present and in his obsessive pursuit of the future, he is in a tradition of American protagonists, all of whom stem from this frontier ethic. In particular, he resembles Fitzgerald's Gatsby in the lack of true culture that makes Gatsby unable to tell illusion from reality, so that he relentlessly and self-destructively chases the future as represented by that green light at the end of Daisy's dock.

At the conclusion of *The Great Gatsby,* Fitzgerald links Gatsby's quest to the American Dream that has lured men like a siren from the first sight of the New World. As he gazes at the Long Island shore, the narrator, Nick Carraway, meditates, "And as the moon rose higher the inessential houses began to melt away until gradually I became aware of the old island here that flowered once for Dutch sailor's eyes—a fresh, green breast of the new world." Fitzgerald concludes that although the dream has elements of nobility, it is ultimately futile and prevents Gatsby from coming to terms with his past. "He had come a long way to this blue lawn, and his dream must have seemed so close that he could hardly fail to grasp it. He did not know that it was already behind him, somewhere back in that vast obscurity beyond the city, where the dark fields of the republic rolled on against the night."

So, too, is Rion Outlaw destroyed by his dream. At the conclusion of the novel, he is alone; he has killed his brother and his wife is on her deathbed. After all this destruction, he begins to realize the futility of his quest.

> And he himself as soon as he had grown to manhood had looked at the mountains and could not rest until he knew what lay beyond them. But it seemed that a man had to flee farther each time and leave more behind him and when he got to the new place he looked up and saw Orion fixed upon his burning wheel, always pursuing the bull but never making the kill. Did Orion will any longer the westward chase? No more than himself. Like the mighty hunter he had lost himself in the turning. Before him lay the empty west, behind him the loved things of which he was made (469).

In this tragically lovely passage, Caroline Gordon universalizes her theme of masculine restlessness by turning to the constellation Orion on which men have projected their quests for centuries. For Caroline Gordon, the American Dream is just another formulation of this eternal masculine ambition. The epigraph to Part II of *Green Centuries,* from Thucydides, speaks to this universal question, to the American pioneering spirit, and to restless, roving Allen Tate, unable to remain at Benfolly: "and as they thought that they might anywhere obtain their necessary daily sustenance, they made little difficulty of removing: and for this cause they were not strong, either in greatness of cities or other resources."

In a short story that appeared in the *Southern Review* in the spring of 1939, Caroline Gordon explores the sensibility of one of Rion Outlaw's spiritual descendants. Like her characters in *The Garden of Adonis* and the end of *Penhally,* the narrator of "Frankie and Thomas and Bud Avery" is a struggling member of the land-poor Southern gentry. His true poverty is inward, however, since he is a man blinded by pride and avarice, as Gordon's first paragraph reveals. "The first year I was at Taylor's Grove, I raised ten thousand pounds of tobacco. Five thousand of lugs and seconds and five thousand pounds of prime leaf. And, boy, was it prime!"[5] When a white drunkard tries to seduce the wife of the narrator's fine black tenant, the narrator blames the wife. The first paragraph concludes, "I ought to have got thirty cents for that leaf, the way it was selling that year. But I didn't get but fifteen. That yellow wife of Tom Doty's was the cause of that."

For her first collection of short stories, *The Forest of the South* (1945), Caroline changed the title from "Frankie and Thomas and Bud Avery" to "Her Quaint Honour," a phrase from Andrew Marvell's "To His Coy Mistress." Although the change in title may be attributed to the vogue for Marvell's poem after its New Critical explication in Cleanth Brooks and Robert Penn Warren's *Understanding Poetry* (1938), the alteration also underlines the story's themes. The "quaint honour" of the title is a double-edged irony. The narrator would be completely incapable of appreciating the witty conceits of Marvell's "To His Coy Mistress," but he would probably agree with its cynical sentiment that chastity is not worth preserving in this world, particularly in this case where the honor at issue is that of a black woman and loss of profit is involved.

Caroline was not the only productive member of the Tate family during their time in Greensboro. As Caroline found inspiration for *Green Centuries* in the Cherokee National Forest, so did Allen's fishing excursions there provide the source for one of his best poems, "The Trout Map" (1939). After a four-year hiatus, he discovered that he was able to write poetry again.

While teaching at the Women's College, Allen also managed to finish his Civil War novel, *The Fathers* (1938), on which he had been working, in various forms, since 1932. Like Caroline, he had been exploring the sources of modern evils and, true to his Southern credo, found some of them in the Civil War. Ultimately, like his friend Red Warren, he traced present evils to man's nature and the workings of original sin throughout history. These answers, however, only led to further, unanswerable questions, such as those asked by Lacy Buchan, Tate's narrator: "Why cannot life change without tangling the lives of innocent persons? Why do innocent persons cease their innocence and become violent and evil in themselves that such great changes may take place?"[6]

The Tates did not spend all their North Carolina years pondering such weighty questions. True to their gregarious natures, they traveled to make visits and were visited, attended parties and gave them. In Nashville, they kept in touch with the Agrarian brethren and met Andrew Lytle's bride, Edna, with whom they were quite taken. In the summer of 1938, the Tates rented a house in Cornwall, Connecticut, near the family of critic and poet Mark Van Doren who was at work on his *Shakespeare.* In his autobiography, Van Doren recalled how vividly Allen, in the process of completing *The Fathers,* could bring the past alive for others: "none of us has forgotten how Allen set up the [Van Doren] boys' blackboard on our lawn and diagrammed the battle of Gettysburg, explaining every move the armies made."[7]

In Connecticut, Caroline seemed more preoccupied with the present, a new dachshund, Bibi, and new acquaintances, such as the young poet John Berryman, of whom she wrote to Robert Lowell, "We were talking about [John Crowe Ransom's poem] The Equilibrists and when I mentioned several lines I particularly liked Berryman recited the whole poem and followed up with the Antique Harvesters and Captain Carpenter. (We have no books in this house except Abolitionist tracts and botanical works so a young man with a memory like that is handy to have around.)"[8] She did not, however, forget old friends, and during their New England summer the Tates managed a trip to Cape Cod to visit the John Peale Bishops, as well as Edmund Wilson and his wife, the novelist Mary McCarthy.

One reason these sojourns away from home were so appealing to the Tates is that they relieved Caroline from the burdens of housekeeping. Allen wrote to an organizer of the Savannah Writers' Conference that Caroline's real, time-consuming work was her pursuit of decent household help; writing was her leisure activity.[9] In addition to relief from domesticity, writers' conferences provide the additional bonus that authors are paid to discuss their work with their friends. The Tates could have the benefits of the Benfolly ménage without its burdens, and they would attend countless writers' conferences in years to come.[10]

At Savannah they were reunited with Andrew Lytle and John Peale Bishop. Caroline spoke on the "Southern Short Story" and used her story "Old Red" to illustrate technique in a subsequent talk. Tate lectured on writing a novel and the analysis of a poem. Biography and magazine writers were covered by Lytle, while Bishop discussed propaganda on one day and myth on the next.

As the program of the Savannah conference illustrates, at this time Caroline was regarded as more or less Allen's equal as a writer. At the Women's College, this was also the case since they both held the rank of professor and had equal teaching loads. In later years, Caroline regarded this situation with longing and bitter regret since she was soon to be demoted to the rank of faculty wife; similarly, her reputation would decline while Allen's skyrocketed during the 1940s.

In the spring of 1939, Allen could not foresee the consequences for Caroline and the concomitant strain on their marriage. His reasons for a move seemed sound. After some testing of the waters at Southern schools, Allen decided to accept Princeton's offer to initiate and direct a program in Creative Writing. He did not feel that the Tates were appreciated at the Women's College, except for their publicity value; at Princeton, he would teach only writing which he preferred to teaching literature; and, finally, Princeton was offering him such a good salary that Caroline could devote herself to her writing since women did not teach at Princeton.[11]

After a summer at Monteagle, Tennessee, with Andrew and Edna Lytle, in the fall of 1939 the Tates moved to 16 Linden Lane in Princeton where they would remain until the summer of 1942. Despite Allen's mockery of academia in his poem "The Ivory Tower" (1936), he may have immured his family in the quintessential ivory tower when he joined Princeton. The university, with its Gothic architecture and beautiful landscaping, was nestled in the heart of a quiet but elegant small town. Because of its proximity to New York City, less than an hour away by train, Princeton was an attractive residence for wealthy businessmen as well as impecunious professors, providing an unusual social mixture in which Caroline, as a novelist, delighted.

Both Allen and Nancy established themselves quite rapidly. He found his teaching acceptable but wrote of his students, with some Agrarian snobbery, that what they knew about life was all that could be expected from commuters' heirs.[12] Ironically, soon he too was commuting regularly to New York where he was a panelist on "Invitation to Learning," moderated by Huntington Cairns, in 1940 and 1941. Tate's literary biographer, Radcliffe Squires, provides an account of the program and Tate's participation. "Each program . . . concentrated on a particular literary work—*The Divine Comedy, The Turn of the Screw,* and so forth. The colloquies with participants such as Mark Van Doren, Katherine Anne Porter, John Peale Bishop, and Jacques Barzun were superb . . . Some of the discussions were published in 1941 in *Invitation to Learning* edited by Tate, Huntington Cairns, and Mark Van Doren. . . . Amusing, generous but cunning, he plays those who disagree with him as a good fly fisherman plays a trout he respects."[13]

Nancy was as successful as her father in her own kind of fishing. Now fourteen, the plump and amusing child had developed into a devastatingly beautiful teenager with a predictable effect on Princeton undergraduates. Caroline, who was considered plain in her youth, ruefully wrote to Malcolm Cowley's wife Muriel, "I would have thought it would be fun to have a daughter who just mowed the men down but it keeps us worried stiff—she is so headstrong and so precocious. I think poet's daughters are likely to be that way."

While husband and daughter were succeeding in their own ways, Caroline found herself classified as a faculty wife while she thought of herself as an author, at work on *Green Centuries.* Behind her usual epistolary mask of comedy, she expressed her feelings to Muriel Cowley:

> Must get back to my Indians. I plan to kill off twenty six of them today but alas, I will have to stop the bloody work at four o'clock to go and pour tea at a ladies' gathering. The faculty ladies here are all great organizers—and callers. If they know who you are they call on you because they like to call on writers. If they don't know who you are they call more than ever to console you for being so obscure. I am known in the town as a writer of mystery stories under a nom de plume. I hope they think I am Mignon C. Eberhart [a popular novelist].[14]

An omniscient being in the world of her novel, she found herself powerless and "obscure" in the somewhat rigid and narrowly focused hierarchy of Princeton's patriarchal society.

Caroline responded by finding some congenial friends of her own within and without Princeton "society." Within that society was Grace Lambert, a distant cousin, and the wife of a very wealthy man, Gerald Lambert. At the Lambert's residence, which Caroline considered "the best rich American house that I ever saw," Caroline was given the "mushroom rights" to the grounds so she could pursue her passion for mushroom hunting. Because the Lamberts had servants, Caroline could indulge in a two-day "rest cure" with Grace Lambert in which the women stayed in adjoining beds, talking and resting. A "rest cure" was Caroline's way of withdrawing from her responsibilities when they threatened to overwhelm her.[15] She would often take them with women friends, like Muriel Cowley who spent such a day with her, sipping whiskey and talking, while Malcolm and Allen attended the Harvard–Princeton game in the rain.[16]

Because of Princeton's central location on the New York–Washington corridor, Caroline could also keep in contact with old friends such as the Cowleys in Connecticut, the Edmund Wilsons and John Peale Bishops on the Cape, and her cousin "Little May" Morse in upstate New York. Many visitors would simply stop by on their way to somewhere else, causing an impromptu party, often with much drinking, Caroline's excellent cooking, and a game of charades. In the summer of 1940, the Tates went to Vermont to attend the Bread Loaf Writers' Conference where they saw two Agrarian brethren, Donald Davidson and Andrew Lytle.

Andrew Lytle was a close friend in these years, as important to Caroline as he was to Allen. In the late summer of 1941, the Tates shared the Lytles' "Log Cabin" at Monteagle. There Caroline could realize a way of life close to her Kentucky childhood. She enjoyed a communal household, both for its society and its ability to serve as a buffer in her sometimes intense and tense relations with Allen. Both in its lack of modern conveniences and its plentitude of country food and cheap labor, life at the Log Cabin reminded her of grandmother's farm, Merimont. "The kitchen is like a cave, with its old fashioned dirt floor. You communicate with the servants—when there are any by yelling down a dumb waiter and everything is pretty rough but Andrew's father comes up from the plantation every week end with corn, tomatoes, beans, melons—and fresh negroes if they are wanted. And it is just the kind of life we both like."[17] Since she had recently finished correcting the proofs of *Green Centuries,* she was free to indulge herself by roving in nature, as she had in her childhood, while hunting mushrooms.

In addition to their shared proclivity for this way of life, Caroline and Andrew shared a vocation as novelists. Andrew was in the last stages of a novel, *At the Moon's Inn* (1941) and Caroline enjoyed discussing her craft with him, as she had when they had worked together at Monteagle in 1937, Caroline on *None Shall Look Back* and Andrew on *Bedford Forrest and His Critter Company.* Further, Lytle was her kind of writer. Some of the comments Caroline made about him for Bobbs-Merrill publicity material apply equally well to herself. "Lytle has the Flaubertian characteristic. For him a period of antiquity is a country which he must enter and travel about in; more than that a world which he passionately longs to apprehend. It is this reverence for the particular detail, this passionate apprehension of a whole world that gives Lytle's writing its peculiar power."[18]

With such respites as the sojourn in Monteagle, the Tates continued their life in Princeton. Caroline found resources in observing the great and encouraging the obscure. In the first category was the poet Wallace Stevens, who came to lecture in the spring of 1941. Caroline called him "the queerest fish" and was fascinated by his "double life." "He is one of the best poets writing in the world today and is also a big business man, vice president of some big insurance company. We marvelled at him and felt that he was something of a monster. . . . The strain seems to be getting him down. He says he doesn't sleep but four hours a night."[19] Although Allen might well marvel at Stevens' ability to write poetry and earn a substantial living, Caroline seemed interested in the "strain" of a double life, which makes you a "queer fish" or "something of a monster." She may well have been questioning her many roles as housekeeper, wife, mother, and author and wondering if she, too, should restrict herself to four hours sleep to keep up with it all.

If she must look up to Stevens, she could feel herself above William Slater Brown, a protégé of her first years at Princeton.

Brown was an aspiring novelist who had been E. E. Cummings' companion in the enormous room, husband of Allen's former colleague at *Telling Tales,* Susan Jenkins Brown, and the Tates' neighbor when they lived in Patterson, New York. Now suffering from alcoholism, he moved in with the Tates for several months, starting in the spring of 1940, so that Caroline could supervise him and the writing of his novel.

Although Brown's novel progressed, Caroline's watchful eye could not prevent him from falling in love and having an affair with Helen Blackmur, a painter and the wife of poet and critic R. P. Blackmur. The situation was further complicated by the fact that Richard Blackmur was Allen's assistant in the Creative Writing Program.[20] To Caroline's dismay, Bill left town. She wrote to his editor at Bobbs-Merrill, "I am much distressed that we can't have Bill here while in the throes of finishing the novel, but the lady he is in love with is still in town. I fear it would be fatal for him to be in the same town with her just now. If Allen will let me I will go up to his place and stay a few days and lend a hand toward the last."[21]

Caroline did not join him, and Bill went on a binge. Caroline wrote to the Cowleys, his neighbors in the Tory Valley, to ask them to keep an eye on him and set up an account at the local grocer's that Caroline would control. She wanted Bill to return to Princeton, but "It looks hopeless but I hate to admit that it is. Allen feels that it is, which is one reason he won't let me take him back to Princeton." Allen's other reasons may have involved Bill's drinking behavior or the constant presence of an outsider in the inner world of his home. There is no suggestion of any romantic interest here, and Caroline was notoriously chaste.

This curious incident sets a pattern that would continue for the rest of the Tates' marriage. As Caroline's reputation slid into comparative obscurity while Allen's prospered, she increasingly bolstered her sense of self-worth by taking in "lame ducks" of one kind or another. At the time, she would have a high opinon of the "lame duck's" unrealized potential. She told Brown's editor at Bobbs-Merrill that Brown was "the writer who has been absent from the contemporary scene for a long time—the major novelist." Whatever temporary balm such a protégé gave to her wounded ego, it only added strain to her marriage; Nancy Tate Wood remembers her father asking her mother why they always had to have someone living with them.[22]

Caroline's mood darkened further in the winter of 1941–1942 by the response to *Green Centuries* which had been published by Scribner's in the fall of 1941. She wrote to Ward Dorrance, a young writer, "My book was a complete failure, financially, didn't even pay back its advance. . . . It was poor timing, of course—people are tired of pioneer stories, and I think that in a way it is a hard book to read, but I had expected it to do better than it did."[23] Her disappointment was exacerbated to wrath and envy by a trip to New York in early November, as she explained in a letter to Mark Van Doren.

> I did not see my book on display in any book store, and finally in a very vicious mood, after looking at the Saratoga trunk [a popular novel] all over the place, walked into Scribner's book store. I could not find my book anywhere so asked a clerk if he had a book called Green Centuries. He led me to a table, piled high with garden books and so help me God proferred me a herb manual. This brought my already fevered blood [to a] boiling point. I flounced upstairs, got hold of Bill Weber and scared him to the point where he is willing to let me have some say about the advertizing. . . .
>
> By the way, speaking in a malicious vein, and there is not room for anything much but malice in my nature this morning, I was delighted to see that Dorothy [Van Doren] told Martha Gellhorn [writer and third wife of Ernest Hemingway] off. . . . Imitation is an amiable trait in a wife but not so good in a professional writer and those stories [*The Heart of Another*] are so patently imitative that it is embarrassing to read them. I suppose what started me off on my Put Gordon over campaign was opening the Sunday book reviews last week and seeing myself decently veiled in black (Scott Fitzgerald is dead and all of us Scribner authors are in mourning [a reference to the posthumous publication of Fitzgerald's unfinished novel *The Last Tycoon*] while Miss G. had an ad all to herself. Yes, that worked like madness in my veins. And I have been on the rampage ever since. Well, you know hell hath no fury like a woman (writer) scorned.[24]

Although relegation to the gardening section and insufficient advertising might make any author's blood boil, this letter also is illustrative of Caroline's characteristic response to lack of success in the 1940s. She becomes angry, briefly at Scribner's for not promoting her enough, but then at another woman writer, in this case, Martha Gellhorn, whom she perceives as a rival. She does not see herself in competition with F. Scott Fitzgerald or Ernest Hemingway, but only measures herself against the treatment of another woman, as she would later with Katherine Anne Porter. The knife was given another twist here by the fact that she perceives Hemingway's reputation as helping his wife, whatever her true literary merit, while Caroline found herself overshadowed by Allen's prominence, not nurtured by it.

Caroline's sense of her world gone awry was also reinforced by the world going awry with the entry of the United States into the Second World War. She wrote to Ward Dorrance, "I feel as if some horrible Grendel were lurking in the marshes, bellowing for a sacrifice of young men, and that all our business, nowadays will be to pack them up and ship them off to him properly." She was also sickened by what she perceived as the patriotic posing and emotional self-indulgence the war seemed to evoke in noncombatants: "it isn't so much the sacrifice—I am not a Pacifist and I know that war is spiritually necessary—but the fol de rol, the asininity, the hypocrisy that accompanies it." By "spiritually necessary," she means her belief that a hero can only fulfill his potential in a struggle with death, a concept well-illustrated in her fiction, particularly *None Shall Look Back* (1937) and *The Glory of Hera* (1972).

The war also seemed to mirror and augment the Tates' domestic uncertainties. In the spring of 1942, Allen knew that Princeton would not extend his contract. He believed he had been displaced by his assistant R. P. Blackmur. Russell Fraser, Blackmur's biographer, heard the story from Allen. "He [Blackmur] 'connived' was the way Allen put it. Richard said he expected to depend on Tate for 'orders and aids.' But he had to dispose of his rivals, Allen said, and the deference was only a blind. . . . When Jacques Barzun in a book review called Allen a Fascist, Richard went to [Princeton Dean Christian] Gauss in defense of his friend. 'Allen isn't *really* a Fascist,' he said."[25] Whatever the explanation, Allen was out of a job.

Allen toyed with the idea of joining the Navy, but after he was rejected due to a stretched muscle in his shoulder, he seemed to lose interest in the idea, much to Caroline's relief. "There is no telling what sort of work he will be put at if he does get in and then it is impossible to get out during war except by a dishonorable discharge," she wrote to a cousin.[26] Caroline explored a job at Sarah Lawrence but said "they were so insulting to Allen that I withdrew my application."[27] Louisiana State University seemed to rush to the rescue with an offer for Allen of the job Robert Penn Warren had just relinquished to move to the University of Minnesota. The Illinois Institute of Technology also made him an offer.[28] Her husband, however, had other ideas, as Caroline wrote to Mark Van Doren. "Allen says he has a novel in his head now and isn't going to take the job. We figured out, too, that with our combined advances we would have as much money as the job would bring, more when we consider the difference in living expenses at Monteagle and Baton Rouge. . . . I'd rather go to Florida but Allen says he can't work in a new place and he has never been able to."[29]

Although Caroline did not get to spend the year in Florida, the move to Monteagle made her a more equal marital partner. Since Allen no longer had a salary, each contributed what they had, their advances; they had returned to the status quo ante Princeton. Another sign of a fresh start was the return of Caroline's creative impulse despite her discouragement over the failure of *Green Centuries.* She described the new novel in a letter to Ward Dorrance. "I'm calling it The Women on the Porch. The porch is affixed to an old house like Merry Mont. The porch is a sort of *stoa* to Hades. If the story has any form it is that of a myth, Eurydice and Orpheus." Unlike *Green Centuries,* it would be set in the present and would not entangle her in the time-consuming research she loved. She merely went to New York City "to do in one morning about all the research for my new novel: my hero is a professor of history at Columbia and I had to find him an office so that I'd know what he looked out at, how the light fell, etc."

Most of the novel does not take place in New York, but at Swan's Quarter, a thinly veiled Merimont, where Catherine Chapman has fled after learning of her husband's infidelity. The house, the farm, her grandmother, and her marriage are in a state of imminent collapse. Further, Swan's Quarter is a world of women in retreat from men. Her cousin Daphne was abandoned by her husband on her wedding night, her Aunt Willy is a spinster, and her increasingly senile widowed grandmother lost the support of her only son because he broke his neck while drunk. This bleak and sterile world is hardly a feminist retreat where women develop their strengths unmolested by men.

Instead Gordon seems to be suggesting that despite their selfish pursuits, men are necessary to women. Women who isolate themselves from men by regressing to a childish state of presexuality stagnate and fester. Catherine muses:

> For a moment she was overcome with compassion. It seemed to her that everybody, that she herself, was like Daphne, half-crushed by some early misfortune and having to advance, maimed, through life, cutting in the very struggle to maintain balance a ridiculous figure that somewhere must provoke mirth. And then she felt a revulsion against the woman, against the very house that harbored her. It is this place, she thought. There was always something wrong with it. And in the silence she listened and heard the sound of the creek and it seemed to her that in its babble it announced its purpose of flowing in a great circle about the farm, of cutting it off from the rest of the world.[30]

Despite this realization, Catherine tries an alternative retreat, an affair with her cousin, Tom Manigault, a neighboring farmer, and even considers marrying him. "There would be the succession of country pleasures which so absorbed and delighted her. It is the life I was made for, she thought, the life I have always missed."

Catherine, however, chooses her husband and New York, even though, or perhaps because, he tried to strangle her when learning of her adultery. This seemingly bizarre decision announces the increasingly diminished stature of women in Gordon's fiction. While it may be practically impossible to live with a man, it is impossible to live without him because Gordon's women, while intuitive and emotional, lack the masculine intellect which she considers superior. She would later find this sentiment encapsulated in a line from the theologian Jacques Maritain, which she used as the epigraph to one of her later novels, *The Malefactors* (1956): "It is for Adam to interpret the voices which Eve hears." That is exactly what historian Jim Chapman does in the last lines of the novel when Catherine and Jim learn that Willy's prize stallion, "Red," has accidentally been electrocuted. "We will bury him as soon as it's light. Then we must go," Chapman declares to the silent and subdued Catherine. Bury the past and go on to the future is his interpretation of the voices of Swan's Quarters.

In autobiographical terms, *The Women on the Porch* (1944) suggests the ways in which Caroline has come to terms with her own past. She knows the pastoral utopia of her childhood summers at Merimont is no longer accessible to her. Indeed, she questions what she learned there, and demonstrated in her earlier novels, about women's courage in the face of masculine abandonment and betrayal. Now she seems to be blaming the victim, inadequate Catherine, as she blamed herself for Allen's infidelity both in her 1933 letters to Sally Wood and in *Green Centuries*.

In Catherine's obedience to the voice of Jim's superior intellect, Caroline seems to be relinquishing her own aspiration to artistic equality with Allen. She may be telling herself that her waning success in the face of Allen's waxing reputation may simply be a result of gender: not just social conditioning, but an intrinsic difference. Both get the raw material, Eve's "voices," but "it is for Adam [Allen] to interpret" them. Willy's destroyed horse is named "Red"; Caroline's most acclaimed short story was and is "Old Red." Over the next thirty years Caroline would write only three more novels and a handful of short stories. They are not her best work.

As Allen and Caroline moved from Princeton to Monteagle in the early summer of 1942, they buried some of their past on the way to the future. Caroline wrote to Malcolm Cowley, the witness to Allen's affair with her cousin in 1933, "We went by Benfolly on our way to Monteagle. It looked mighty pretty with the hollyhocks in bloom but we hardened out hearts and listed it with a real estate agent. It is a white elephant, that house, and we have finally realized it." In a letter to Ward Dorrance, Caroline attributes the attempt to sell Benfolly to Allen's poetic drought during the Benfolly years. "Allen knocked out for years after we acquired Benfolly and even now I hesitate to go there. He never seems to get anything done while he is there."

Despite the fact that Caroline had been able to work at Benfolly, Allen's affair and writing block poisoned Caroline's native region for both of them. The Clarksville area was so inextricably associated with Caroline's artistic and emotional selves that when she relinquished it, she believed she was surrendering the possibility of a permanent home. She wrote that at Foster's Falls at Monteagle, she comforted herself with nature's permanence. "I stood there and watched the water coming over and felt damned good to think that no matter what strange places I have to live in that water is still coming over those falls."[31]

Like Merimont and Benfolly, Monteagle's cottages were characterized by porches, but the Tates did not associate them with the *stoa* to Hades. Monteagle was a place where the Tates had been happy and productive amid congenial companions, and they anticipated more of the same. Their arrival in June was auspicious. Allen, Caroline, and Nancy stopped at nearby Sewanee where a dance was in progress and they stayed up most of the night with the Lytles. Soon Caroline's brother Morris arrived for a visit at Monteagle, and Caroline took up her favored pastimes of swimming and mushroom hunting while Allen turned to swimming and fishing.

Further remembrances of good times past arrived with Robert Lowell who wished to work and live with the Tates as he had in the summer of 1937. This time, though, he was not to live alone in a tent on the lawn, but shared the Tates' house with his wife, Jean Stafford. At work on her first novel, *Boston Adventure,* Stafford would provide a companion and protégé for Caroline. Another significant change was Lowell's recent conversion to Catholicism. As Caroline reported to Malcolm Cowley, "His great worry about coming down here was whether he could find a Catholic church near here. He found one easily."

Stafford and Lowell acted as the buffers that Caroline seemed to need when constantly living and working in the same house with Allen. Nancy was now sixteen years old and happily busy with her own friends and interests at her school in nearby Sewanee. Although Allen seemed quite amenable to the Lowells joining them, Caroline was the one who urged them to come and made all the arrangements. The house, which Caroline called "New Wormwood," seemed ideal for the arrangement. "A spacious and hideous Victorian cottage," it actually contained two apartments with separate entrances, but Caroline wanted a more shared lifestyle.[32] She argued that since only one apartment, the Tates', had a modern kitchen and large living and dining rooms, the Lowells should share these amenities in return for a small workroom for Allen in the Lowells' apartment.[33]

The arrangement seemed to have worked, for the year was a happy and productive one for all four writers. By March 1943 Lowell had completed sixteen poems, the core of his first volume, *Land of Unlikeness*.[34] Stafford and Gordon made great progress on their novels, but the other aspiring novelist, Tate, found himself blocked, and never did complete this novel. Instead, he found himself writing poetry again, and had completed seven poems by January 1943.[35]

On a happy inspiration, Allen also translated the *Pervigilium Veneris,* with its haunting refrain, "Tomorrow may loveless, may tomorrow make love."[36] In his Introduction to the poem, he attributes to Caroline the first line of stanza XXI, "Now the tall swans with hoarse cries thrash the lake." With ironic aptitude, the stanza concerns a raped maiden who is turned into a bird to escape her tormentor, and now "bewails/her act of darkness with the barbarous king." The succeeding, and final, verse, however, speaks more to Allen and his poetic droughts. "She sings, we are silent. When will my spring come? . . . /Silent, I lost the muse. Return, Apollo!" The muse had returned to Allen in this productive period, the last one of his poetic career. Hereafter her visits would be infrequent and short.

Predictably, after Lowell and Stafford left Monteagle, the Tate household also began to fall apart. Both Caroline and Allen were ailing during the early summer of 1943, with what sounds like exhaustion after a spate of visitors. Caroline's brother Morris arrived as well as her Aunt Piedie. From Greensboro came Caroline's former pupil James Ross, and his sister Eleanor Taylor, also a writer, with her husband, the novelist Peter Taylor. Wrote Caroline, once again, "I don't believe I ever worked harder in my life. Just getting enough food for them all kept me hopping." She and Allen started taking "liver" pills in an attempt to build themselves up, and soon they felt well enough to face some new challenges.[37]

Allen had accepted a post for which he had been negotiating for several years, that of poetry consultant at the Library of Congress. In comparison with teaching, his duties would be light and his prestige great. He would inaugurate a series of poetry readings and supervise the compilation of a checklist that became *Recent American Poetry and Poetic Criticism* (1943). Further, the move would bring the Tates to Washington, which during these war years was the hub of activity and a complete contrast to remote Monteagle. The move brought advantages to Caroline as well. Once again Allen, Caroline, and Nancy would share a house with another couple, this time fellow Tennesseeans, Brainard and Frances Cheney who had been introduced to the Tates by Robert Penn Warren back at Benfolly. Brainard, or "Lon," was an aspiring novelist who was then serving as a senatorial assistant. "Fanny" Cheney was a librarian whom Allen had chosen as his assistant at the Library of Congress. After some travels in Connecticut and New York, in August 1943 the Tates rented a house in southern Washington across the Anacostia River. When the Cheneys arrived at the end of the month, Caroline named the house the Birdcage, after a whorehouse in one of Cheney's novels, although others claimed the name was a tribute to all the literary "singing" that would emerge from the dwelling.[38]

Fanny Cheney, however, remembers the name as suiting a small house that did not match its architectural pretensions. The Tates took the upstairs while the Cheneys were below. Fortunately, the birds in their little nest did agree, largely because of Fanny Cheney's capable and amiable nature. At the Library she was invaluable to Allen since she handled most of the minutiae of the job and even signed him in when he wanted to arrive late.[39] At the Birdcage, she was a godsend to Caroline. Fanny not only took over most of the household chores, but aided Caroline with bits of research and advice for her novel. Caroline wrote to Jean Stafford, "I have been floating like a cloud above household cares. . . . Living with Fanny is very demoralizing. She does, without grumbling or apparent effort, all the things that I am accustomed to do with clenched teeth and curses."[40]

Fanny's aid was most opportune, for Caroline was in the final stages of *The Women on the Porch*. She finished the first draft early in the fall, but was dissatisfied with the chapters introducing Jim Chapman and rewrote them twice. Allen agreed that they were "perfectly foul," so in mid-December Caroline tried again. "The other day I hauled off, took a fresh start and wrote a substitute chapter, which, according to Allen, had exactly the same faults as the original."[41] She did not say what these faults were, but they may have arisen from the difficulty of presenting the abstracted and adulterous Chapman's point of view without making him seem so hopelessly unsympathetic that Catherine's return to him would lack verisimilitude.

Late that fall, Caroline's happy life in the Birdcage would be interrupted by a number of events, all in some way connected with the illnesses of others. At Allen's urging, John Peale Bishop had accepted an appointment as a consultant at the Library of Congress, although it meant leaving his family on Cape Cod where his wife was busy with war work. He rented a room across the street from the Tates, but took his meals with them at the Birdcage. Caroline told Jean Stafford that when Bishop developed a bad cold, "everybody—except hard-hearted me—was afraid he'd get pneumonia so we turned Nancy out of her bed and installed him there, and he gasped and wheezed for several days and finally got strong enough to move" to a friends' apartment. She explained her "hard-heartedness": "Allen feels awfully sorry for him but his plight leaves me pretty cold. I am fond of John but he always did make me tired."[42] Caroline was always ready to nurse her friends and relations, whether physically, emotionally, or artistically, but one of Allen's rich and well-known fellow poets simply left her cold. Bishop's health continued to deteriorate, and he left Washington for his home on Cape Cod.

Her nursing skills were soon in demand again, this time for her critically ill father. The octogenarian James Gordon had been living in a boardinghouse in Leesburg, Florida, where he was pursuing his avocation of fishing with his usual passion. He was hospitalized in Leesburg for an enlarged prostate and resultant kidney blockage, but was turned out of that hospital, so Caroline claimed, for slapping a nurse. By the time she reached him in late December 1943, he was in the Orange General Hospital in Orlando, Florida, where he was failing rapidly. His prostate and kidney troubles had been relieved by the insertion of a catheter, but his heart was weak.

Caroline took a room near the hospital and spent most of her time with her father. She described his suffering, physical and mental, as "terrible." He often hallucinated, mainly about animals and hunting. Above his bed, he saw animals, first a lamb, then a pole-cat, so that "his ambition was to get up to the ceiling where all those animals are." On another occasion, Caroline wrote to Fanny and Lon Cheney, he was "obsessed to get over in a corner, through an imaginary gate." All Caroline could do for him was try to make him comfortable and feed him, although she questioned the benefits of her efforts. "I think he would have died, by the comparatively merciful process of starving to death, if I had not got here when I did." Toward the end Caroline kept her sanity with her work, revising *The Women on the Porch* "by hand, standing at the dresser between vomitings."[43]

Caroline had left the Birdcage in the capable hands of Fanny Cheney, so she did not seem worried by her month-long stay in Florida. She did, however, miss a major event. On January 3, 1944, Nancy Meriwether Tate married Percy Hoxie Wood, Jr., a handsome young man from Memphis, currently enlisted in the Navy. Caroline had opposed the marriage, thinking that at eighteen Nancy was simply too young. Allen, as Fanny Cheney recalled, thought otherwise, and he "connived" the marriage while Caroline was away. Caroline seemed resigned to the news, perhaps wanting to accept what she might have felt compelled to oppose if she had been on the scene. She wrote to Allen, "I suppose Nancy + Percy are married by this time. I hope we will have enough money to buy Nancy some silver, but probably not, the way things look."[44] Despite her initial opposition, Caroline became very fond of her new son-in-law, who was to be a mainstay of her final years.

Though the Birdcage may have lost Nancy, it soon gained another inmate, Katherine Anne Porter, who arrived on January 21, 1944, to replace John Peale Bishop at the Library of Congress. She also replaced Bishop as resident invalid since she soon developed pneumonia which confined her to her room in the Tates' basement. Caroline's nursing skills were once again in demand, but, as in the case of Bishop, this patient did not evoke Caroline's unadulterated sympathies.[45]

Caroline was still smarting from the lack of response to *Green Centuries* and her lack of status in the literary world at a time when Allen's reputation was at its zenith. Porter, too, was no longer the fellow struggler of their Greenwich Village tenement two decades ago. Despite her comparatively meager output, a number of masterful short stories, her literary stock was also quite high. Porter, although several times married, was childless, and in effect had mainly led the life of a single woman. Caroline seemed to regard Porter and herself as the grasshopper and the ant, both in literary productivity and the duties of domesticity. She wrote to Jean Stafford:

> As for K.A. she is an actress who happens to have a talent for setting down her emotions in felicitous prose. She would walk miles to get you a bouquet of flowers or a jug of wine and present the gift gracefully but you cannot *depend* on her for anything. The very thought of anyone depending on her makes her wild. This is partly because she is always in a crisis herself and partly because of her histrionic gifts. . . . This is just the way she is and she can't help it. It goes along with her talent, a protective bit of colouring that nature has given her. Without it I doubt if she could have done as much work as she has.

Furthermore, Porter was considered a remarkably beautiful woman who could easily charm a roomful of men and women. Caroline, while a great wit and distinctively attractive, could not compete with Porter's siren-like allure. Her household now seemed to revolve around Porter, not herself.

The small Birdcage was certainly not big enough to house two such prima donnas of fiction, and matters came to a head at the dinner table. Nancy Tate Wood recalls that Porter and Gordon quarrelled over the correct version of a jingle about a turkey. Caroline insisted that Katherine Anne's rendition was wrong and that she was so stupid not to know the correct version that she should leave the house immediately.[46] This ludicrously trivial pretext may have masked Caroline's real hurt. Porter's version of reality, her fiction, was accepted, indeed acclaimed in a way that Caroline's was not, and Caroline wanted her version to prevail, at least in her own house.

Porter moved to a friend's house in Georgetown, and with distance came a renewal of cordial relations. Despite the happy ending, the incident illustrates Caroline's typically outraged reaction to her comparative literary obscurity. She did not measure herself against male writers or question the way men ran the literary establishment. Instead, she perceived herself in competition with other women writers for that masculine attention, as in her envy at the greater advertising Martha Gellhorn received at Scribner's and the greater acclaim that Porter seemed to receive from the literary world in general. Her response may be traced to her reliance on male mentors, such as her father, Professor Gay, and Ford Madox Ford. The pattern, while initially constructive, was ultimately destructive, in that the bitterness she displayed at the lack of such attention further alienated the literary patriarchs and precluded sympathetic understanding from other women writers.

Caroline's bitterness during this period did have several constructive outlets, however. One was her attempt to paint. Several years before the move to Washington, she had tried a pastel sketch of her dachshund Bub confronting a turtle in the woods. The result elated her, and also pleased the initially skeptical Allen. In Washington, Caroline took a course that introduced her to the techniques of painting. She enjoyed the class so much that she regretted its end, but she had acquired a skill that would serve her as a pleasure and resource for many years to come. Her paintings have a "primitive" look to them and animals are her principal subjects. She favored peaceable kingdoms and allegorical series about her pets, such as "Bub in Heaven." Some did depict human figures, such as a scenic version of life at Benfolly. In this art as well as her life, she may have preferred animals, perhaps because, as she once said, animals are always different and people always the same.[47]

The observation of human life was not abandoned since Caroline also continued to explore her predicament through her writing. The three short stories she published in this period explore three women's responses, two tragic and one comic, to their frustration in the world presented to them by men.

"Hear the Nightingale Sing" (*Harper's,* June 1945), a revision of one of her earliest stories, "Chain Ball Lightning," concerns a woman left alone to face hostile intruders, in this instance marauding Yankees. In a world full of the hopeless human misery of slavery, Barbara had tried to restrict her affection to animals of whom she is, like Caroline, "overfond": "When she was a little girl and Uncle Joe would bring a team in to plough the garden in spring, she would look at the mules standing with their heads hung, their great dark eyes fixing nothing, and she would think how, like Negroes, they were born into the world for nothing but labor, and her heart would seem to break in her bosom."[48] The incident not only indicates Barbara's obsession with animals, but prepares the reader for her revolt when she is no longer among the pitying privileged, but is one of the underdogs herself.

The man for whom Barbara forms a pitying attachment, Tom Ladd, possesses some of the quiet harmony she sees in animals. "He had what her father called 'the gift of silence.' But sometimes, sitting in company, you would look up and find him watching you and it would seem that he had just said something or was about to say something. But what it was she never knew." His gift to her is as ambiguous as his silence. He presents her with a mule, Lightning, who throws and tramples the Yankee soldier when he tries to steal the animal from Barbara. According to Andrew Lytle, the Yankee "dies by the feet of the brutal instinctual forces released by war, symbolized in the sterile mule. The gift of the mule by Ladd to Barbara stands for the hopelessness of their relationship. It does destroy the enemy, but it can stand only negatively in Ladd's stead."[49]

Right relations between men and women are blighted by the curse of slavery which turns Barbara's affections from humans to animals, and then by the war which removes Tom Ladd. The scene in the parlor in which the Yankee soldier tries to act like a guest, after forcing the women to share their meager supper with him, is an acerbic parody of courtship. He sings a song about the nightingale that Barbara associates with her last meeting with Tom Ladd. The only emotion Barbara can muster is unadulterated hatred and her fantasies involve mutilation, quite a change in so tender-hearted a girl. "She looked away, thinking how you could set your thumbs in the corners of those lips and rend the mouth from side to side and then, grasping in your hands the head—the head that you had severed from the body—you would beat it up and down the boards of the well sweep until you cast it, a battered and bloody pulp, into those grasses that sprung up there beside the well." The violence of Barbara's feeling is in proportion to her frustration and impotence in a world she does not understand and for which she had not been prepared.

In "The Forest of the South" (*Maryland Quarterly,* 1944), the equally powerless Eugenie Mazereau retreats into a vindictive madness, but the story concerns not only masculine insensitivity toward women, but the North's moralistic obtuseness to Southern culture. The central intelligence is Lieutenant Munford, a Union officer who falls in love with the daughter of the house his forces are occupying. He is uncomfortable about courting Eugenie under the circumstances, but never doubts the Northern cause. "He would do it all over again, to strike the shackles from the wrists of slaves."[50] Despite the warnings of his superior officer and Eugenie's betrayal of her own cousin to the Federal soldiers, he persists in blindness to her madness. "He swore that he would make her forget everything that had happened." "Everything" includes the murder of her father and her mother's premature senility. M. E. Bradford points out that "Munford's innocence" is the "product of a mild but ineducable millennialism," and, "in the path of Amasa Delano of Melville's 'Benito Cereno' and of his successors among righteous New England heroes, he never falters in the course of his optimism. Eugenie's mad smile in the Macrae garden is to him only an invitation. That it might be a vengeful snare never occurs to his confident simplicity."[51]

"The Forest of the South" is in many ways a Southern version of Hawthorne's "Rappacini's Daughter" that might better be called "The Garden of the South." The story is full of menacing gardens. The first image is the blowing up of Clifton with its famous gardens by a Yankee engineer who destroys it in a fit of pique. In contrast to the engineer, Munford refuses to recognize the poisonous effect of the exotic Southern garden on a Northerner, although he senses it. As he watches a brilliant hummingbird and hears men's distant voices, he muses that "He had never known it so quiet before. But the stillness was oppressive, and the landscape, he thought suddenly, too bright. This shining air held a menace." In the deserted garden of her cousin's house, he proposes to Eugenie, even though he associates her with the threatening Southern vegetation: "The lids [of her eyes] were heavy, so heavy that they dimmed the brilliance of her glance. And the lids themselves had a peculiar pallor, wax-white, like the petals of the magnolia blossom. When he had first come into this country he had gathered one of those creamy blossoms only to see it turn brown in his grasp." Love once more withers in the poisonous atmosphere of the Garden of the South, and paradise is lost again.

In "All Lovers Love the Spring" (*Mademoiselle,* February 1944), Gordon tells her story in a comic mode through a narrator who is a survivor.[52] Unlike Eugenie Mazereau or Barbara, however, middle-aged Miss Fuqua is a denizen of the New South that precludes both tragic losses and heroic gestures. In every sense of the phrase, she is a lady in reduced circumstances. Like Caroline herself, she fondly remembers her childhood when she lived in a "handsome old brick house" presided over by her father, "quite a learned man." With her brothers and her cousin Roger Tredwell, the hero of her girlhood, she would play in the silver poplars behind the house. Her father is now dead, the house has been burnt down, and Miss Fuqua is left alone to care for her invalid mother. Roger has married a paragon of silliness and become a "booster" of the New South. His unromantic appearance prevents Miss Fuqua from even indulging herself with fantasies of unrequited love. "He had taken on weight since he got middle-aged, and the Tredwells turn bald early. When a man gets those little red veins in his cheeks and his neck gets thick, so that it spreads out over his collar, there is something about a dinner jacket that makes him look like a carp."

Miss Fuqua, however, is one of Caroline's characters who can be a successful lover without a successful union,[53] and perhaps may represent a hope Caroline held for herself in her current discouragement. As a lover, Miss Fuqua loves the spring, but with a realistic appraisal of its place in her life.

> On a mound of earth, in that black, swampy water, a tame pear tree was in bloom. An apple tree will bend to one side or fall if you don't prop it up, and peach trees don't care which ways their boughs go up, but pear branches rise up like wands. Most of the blossoms hadn't unfolded yet; the petals looked like sea shells. I stood under the tree and watched all those festoons of little shells floating up over my head, up, up, into the bluest sky I've ever seen, and wished that I didn't have to go home. Mama's room always smells of camphor. You notice it after you've been out in the fresh air.

Like the pear tree, Miss Fuqua does not need a man to prop her up. Her spirit is still capable of rising like a wand in the forest of the New South. Caroline, however, does not choose to end Miss Fuqua's story on this lyrical note. Miss Fuqua must return to mundane life and Mama's camphor smelling room. Another of Caroline's women characters resumes her secondary but necessary role as caretaker of the endless vicissitudes of everyday life.

The intertwined roles of women and the South that Caroline

explored in these three stories also became the main themes of her first collection of short stories, *The Forest of the South* (1945). The volume includes seventeen stories, all of her published tales with the exception of two that were excerpts from *None Shall Look Back* and *The Garden of Adonis*. One chapter from *Aleck Maury, Sportsman* that appeared as a short story, "The Burning Eyes," is also included in the collection. Excluding the excerpts from her novels, Caroline Gordon published sixteen short stories over a period of sixteen years. She did not rest on this achievement, but arranged her disparate works into a book that tells a story of its own.

The order is approximately chronological. The first story, "The Captive," is set on the Kentucky frontier in 1787. Three Civil War stories follow. In the next five stories Gordon takes Aleck Maury from his Reconstruction childhood to advanced old age in the 1930s, a break in the volume's chronology that promotes characterization. "Tom Rivers" is framed in time-present, the 1930s, but the body of the story is a reminiscence of turn-of-the-century Texas. "The Long Day" and the autobiographical "Summer Dust" appear to be set in Caroline's childhood in the early 1900s. "Mr. Powers" and "Her Quaint Honour" concern problems with tenant farmers in the 1930s in the Clarksville setting of her Benfolly years. The years prior to the Second World War also appear to be the locale for the last three stories, "The Enemies," "The Brilliant Leaves," and "All Lovers Love the Spring." Gordon's intention is clear. As in her first six novels, she is attempting to portray and explain the decline of the South.

The arrangement of the stories also highlights Caroline's obsession with chronic misunderstandings and inadequacies between men and women. Andrew Lytle attributes this friction to historic causes: "the conquest of the south is the destruction of a society formal enough and Christian enough to allow for the right relationships between the sexes."[54] The similarities in the romantic problems throughout the collection, however, argue against Lytle's conclusion. The volume begins and ends with a first-person account of a woman alone in the forest of the South. Despite the fact that the women in "The Captive" and "All Lovers Love the Spring" are separated by 150 years, they share the same plight, the withdrawal of masculine support that forces them to a lonely independence in the precarious world made by men. Jinny Wiley, the pioneer, is kidnapped by Indians while she is alone with her children on her farm while her husband is off doing business, and Miss Fuqua, the spinster, is captured by her invalid mother after Roger Tredwell fails to claim her.

Although the stature of the protagonists diminishes with the South from the tragic to the pathetic, Caroline's profound pessimism about the relations between the sexes remains constant. In this context, the forest of the South becomes more than a geographical setting; it is Caroline's dark metaphor for the human condition. Like the couples in *A Midsummer Night's Dream,* Caroline's lovers are separated and lost in the forest but for them there are no happy reconciliations.

Caroline, in some ways, seems to be writing the script for her own loss of the garden of the South. Over the next two years, from the spring of 1944 to the spring of 1946, she would lose Allen, and with that loss of masculine support the garden of the South would be transformed into the menacing forest of the South for her. She would never live there again.

Aside from Caroline's discouragement over the reception of her work, there was no indication of the approaching stormclouds in the spring of 1944. Allen had been offered the editorship of the *Sewanee Review,* which meant the Tates could return to the area close to Monteagle that held pleasant memories for them both. As the home of the University of the South, Sewanee would provide some congenial companions as well as a lovely rural situation amid the mountains. The Tates rented Benfolly to two U.S.O. hostesses for the duration of the war, and for themselves rented a large house on edge of Sewanee near the woods. Nancy, now six months pregnant, had rejoined her parents while her husband was in the Navy. She went to Sewanee in May to prepare the house where they joined her in July.

As usual, the Tates' initial occupation of a new dwelling was cheerful and auspicious. Caroline was delighted with the house which she said was "simply wonderful," but "too good for us," as if they somehow no longer deserved a stable and stately home. After the cramped Birdcage, a large living room, dining room, butler's pantry, kitchen, library, and five bedrooms must indeed have seemed a bit overwhelming. Caroline was also enchanted with the spacious grounds and resolved to reclaim some of the neglected formal garden. Despite her pleasure, she gave the house a somewhat inauspicious name, the Robert E. Lee, "because it has such noble proportions and also because the protuberance on the front porch makes it look as if it were steaming down the Mississippi."[55] Lee lost the War between the States, and at the Robert E. Lee, both Allen and Caroline would lose the war between the Tates.

Allen and Caroline followed the usual pattern of the tense episodes in their relationship. When Caroline felt she was not appreciated as an artist, she made herself necessary to a circle of dependents. In this instance, she was unhappy about the critics' reception of *The Women on the Porch* which she felt had missed the point of what she had attempted. Injury was added to insult when the novel only made thirteen-dollars profit. Further, she was stuck on an abortive attempt to write a novel about a paratrooper with "Icarian" symbolism. She could distract herself from her pain by immersing herself in the troubles of others and new work of her own.[56]

Unfortunately Allen was not included in her sympathies. Although ostensibly successful as the editor of a prestigious review, he was having his own artistic troubles since he was at the beginning of a poetic drought that would last for several years.[57] He could receive little sympathy from Caroline for she would wall herself off from him with her own misery, her circle of dependents, and her work.

Caroline's first "dependents," though, were happy ones for both Allen and Caroline. On Caroline's forty-eighth birthday, October 6, 1944, Nancy gave birth to a son, Percy Wood III, affectionately known as P-III. Caroline remained with Nancy at the hospital and wrote to her son-in-law that her first grandchild was "as cute as he could be."[58] In November, Nancy left Sewanee to rejoin Percy for as long as he would be stationed in Brookline, Massachusetts, and Caroline, like her own mother with baby Nancy, took care of P-III at home.

Once Nancy had found an apartment in December, P-III moved to Brookline and Caroline and Allen were confronted with the prospect of living alone together. Caroline quickly preempted this possibility by undertaking the care of her fatally ill maternal aunt, Louisa Meriwether, the mother of Caroline's favorite cousin and girlhood companion, "Manny." At first Aunt Loulie was quite active, and attempted to take over the kitchen and run the household. When Percy was assigned to the Pacific at the end of February, Nancy and P-III rejoined her parents, and the tottering Aunt Loulie next tried to manage the nursery so that Caroline was compelled to hover over her aunt as she hovered over the bassinet. For Nancy and Allen, the situation was intensely trying and depressing, but Caroline seemed to thrive on it, and managed to get *The Forest of the South* ready for Scribner's. The Tates' relationship was not equally prosperous, however, and Caroline wrote to Jean Stafford in the spring of the 1945, "I hardly ever exchange a word with Allen."[59]

After Loulie's death, Caroline wrote a short story, "The Olive Garden," which marked a transition in her work and her life.[60] That spring of 1945 she had been reading F. O. Matthiessen on Henry James, and the critic's insights led her back to the Master's works. "The Olive Garden" is quite Jamesian in style, plot, and narration. The first sentences echo James's hesitancies, indirections, and refinements as Gordon introduces the reader to the story's central consciousness:

> On the way over Edward Dabney told people that he intended to stop in Paris. He did not mention the name of the other city. And indeed, he had never actually planned to go there. It was merely that the idea, coupled with an incident from the past, had so persistently presented itself to him all that spring. An incident? The happening had not the stature of an incident. A shadow, rather, that all that spring had seemed to brush his past.

Like the protagonist of James's "The Jolly Corner," Dabney is seeking the memories of a happier time, in his case, the years prior to the Second World War, when he was engaged to Susan Ferris, and both were living in the south of France with another couple, the Matthews.

Dabney finds a postwar wasteland: fountains without water, ruined houses, roads pitted by artillery; it is a landscape that mirrors his interior desolation at the unexplained loss of Susan Ferris. In the gardens of a deserted neighboring villa, Dabney is drawn to, but hesitates to take, Susan's favorite path that leads to a figure of a "cowled Madonna." Although he promises himself he will take that path on his way back, the reader does not know if he actually does, for the story ends with his meditations on the caves below. There, according to legend, Ulysses, Deucalion, and pirates found shelter. In the story's concluding sentence, Dabney finds consolation in the past because he believes it will be repeated: "Far below, in the rocky caves, that would bring forth a new race of men, he could hear the heroes murmuring to each other."

Through Dabney, Caroline is evaluating the Tates' past and indicating the direction of their future. The setting of "The Olive Garden" is Toulon where in 1932 the Tates had spent some happy months with Ford Madox Ford, Janice Biala, and Sally Wood. The blasted wasteland embodies the Tates' marital paradise lost, but through that landscape Caroline is groping for a revivifying change. The path of the cowled Madonna represents her increasing attraction to Catholicism. Dabney's hesitation to take it and his final meditation on heroes, so reminiscent of Allen's "The Mediterranean," uncannily presages Allen's initial reluctance to join Caroline on the way to Mother Church. The story must also have contained powerful resonances for Allen since he praised it highly and published it in the *Sewanee Review*.[61]

As "The Olive Garden" indicates, the consolations of religion were increasingly attractive to Caroline in the midst of what she perceived as artistic and marital stagnation. This tendency was reinforced in a number of ways during the war years. She had seen the calm and sense of purpose Dorothy Day found in Catholicism. The Tates spent the winter of 1942–1943 with Robert Lowell and Jean Stafford in the period of Lowell's greatest fervor, soon after his conversion to Catholicism. She remained in touch with both Lowells and talked with Cal on a trip to New York late in 1944. After her Aunt Loulie's death, Caroline wrote to Stafford, "Tell Cal I am now studying Vedantic philosophy and we can have many a rousing talk about religion. This Vedantic business is wonderful. Fits in with all my intuitive certainties—my feeling about animals, for instance. It is comforting, too, to know where my poor, dear Loulie is at the moment. She is on the moon, enjoying the society of gods." Although her tone is somewhat flippant, she is searching for some sympathetic certainty and wants to "tell Cal" who would understand.

Caroline was also repeating a family pattern of behavior. A number of the Meriwether women turned to almost fanatical religious devotion when thwarted by the world of men, in particular their husbands. When Caroline's own mother felt that the life she desired was constantly blocked by the way the footloose James Gordon wanted to live, she became a zealous Campbellite in her middle years. The family past is not simply a burden, but seems to function as a program, which causes the descendant to design her own life in such a way as to repeat it.

And that was what Caroline was doing in the summer and fall of 1945. Masked by a round of visitors, such as the Willard Thorps from Princeton, the tension was building. Caroline was working on a painting that could hardly have helped matters, "a picture of Allen. It is called 'Thinking Nothing, Nothing, Nothing, all the day,' and shows Allen turning that glazed look that he sometimes turns on his friends on infinity."[62] Nancy was operated on for appendicitis, and Caroline felt that she saw everything through "a mist of fatigue."[63] Finally, the past several years' accumulation of misery came to a head. That fall Caroline and Allen had a violent quarrel in which she chased him around the kitchen with a bread knife.[64] Allen asked her to go, and she did, heading north to leave the forest of the South behind.

CHAPTER 6

Caroline headed north that September of 1945 without any real plans beyond a return to Tennessee in December for the divorce proceedings. Even that event, however, did not seem certain. After she arrived in New York City and took a room in the Hotel Aberdeen, Allen followed to attempt a reconciliation, but, as Caroline wrote to Jean Stafford, he "wrote me at my lodgings that he did not dare to see me and would go on home. All pretty bloody, quite a shambles all around."

In an attempt at normalcy, Caroline conferred with Max Perkins and other editors and tried to keep writing. After ten days in New York, unable to write and eager to leave the furnished room she had taken, she headed for Princeton where Willard and Margaret Thorp found her another furnished room in "a beautiful old house on Boudinot Street." Still restless, she spent a "pleasant but strenuous" weekend with Malcolm and Muriel Cowley in Sherman, Connecticut, and then turned to another pair of old friends, Robert Lowell and Jean Stafford, for help.[1]

The Lowells had recently purchased a house in Damariscotta Mills, Maine, and invited Caroline to spend the winter with them. When their plans changed to a Tennessee winter, the Lowells kept the offer of the house open. Caroline decided to confront Maine alone,[2] but the rigors of a Maine winter were initially more than she anticipated, as she wrote to Ward Dorrance. "The Lowells had left the house in chaos. The only stove that will work is in a room that has no furniture in it." She took refuge with a widow across the street while she organized the chaos. Once she was established in the Lowells' house and in a working routine, Caroline began to look about her and appreciate her surroundings. "This is a perfectly beautiful house. . . . It has a lake at the back of it and a tidal river, the Damariscotta, in front. The village . . . is chock full of beautiful houses, large and small. All around there are wonderful dark hemlock and spruce woods to walk in."[3]

In her isolated northern refuge, Caroline attempted to account for the end of her marriage in letters to friends. At times, she blamed herself for allowing the Sewanee household to revolve around her visiting friends and relations, particularly her dying Aunt Loulie, whom Allen always found hard to take. In a letter to Malcolm Cowley, she proffered another explanation. "In such cases one usually looks for the other woman. In this case there are two: Allen's mother and my Muse. I'd have done better if I hadn't been so absorbed in my own work. But Allen's mother looms larger in the picture. She so tortured him when he was a child that he is literally afraid to commit himself to any woman." Mainly, she attributed the failure to a sort of temporary insanity in Allen, describing him as "possessed" and "completely irrational." She referred to these states as Allen's "seizures."[4] By doing so, she could avoid blaming herself; if Allen no longer wanted to live with her, he was not himself.

Despite Allen's plans to end the marriage, he did not cease looking after Caroline's interests or communicating with her. He wrote to Malcolm Cowley that Caroline should be taking advantage of favorable reviews of her recently published *The Forest of the South* and that he had contacted her agent for that reason.[5] He planned, and Caroline expected, that he would help her find a teaching job. During one period he wrote to her what she described as "daily letters analyzing my character."[6] By late fall, Allen was ready for a face to face colloquy and asked Caroline to meet him in New York.[7]

The talk was apparently successful because by early December she was back in Sewanee with Allen. Although she had returned to him, she was not optimistic about the reconciliation. 'I am not at all sure we will make it," she wrote to Ward Dorrance on December 6. The contradiction between her action and her attitude might have indicated that she needed to prove to herself that the marriage was really over. She left several days later, writing, "We are both cured for good. We realize that we can never stay under the same roof again, and it is something to have that settled."[8]

The divorce was scheduled for early January, and Caroline remained in Nashville for it, at the home of friends, the Alfred Starrs. On January 8, she wrote to Ward Dorrance, "My little affair came up this morning, and was over in fifteen minutes. They do things very well in Tennessee. Allen is in the hospital, not exactly with a broken heart, flu. I find myself better than I have been in months, almost gay. It is such a relief to have that over, to be convinced that there is absolutely nothing I can do for Allen any more, and am at liberty to pick up the pieces of my own life."

Despite the metaphors of recovery from illness or accident she used when discussing her divorce, Caroline was by no means cured, but might be considered among the walking wounded. She decided to establish herself in New York for "business reasons" although she felt "a violent aversion" to seeing Allen's and her mutual friends. Janice Biala found her a cold-water flat in Greenwich Village at 108 Perry Street, and Caroline herself managed to find jobs teaching writing at New York University and reading manuscripts for Macmillan. She needed the routines of work to fill the sudden vacuum in her life. "One has to have some kind of framework and to be suddenly quite alone when one has been used to fighting like a tiger to get a few minutes to one's self, is disconcerting." She tried to see friends every night because "it tires me enough to make me sleep and thus keeps off the horrors."[9]

She even gave her first party without Allen and seemed surprised that "everything seemed to go off just as well as in the old days." Appearances belied reality, as Caroline herself recognized. "I don't think I'm doing well at all. I'm just treading water as best I can. I have to be somewhere—unless I commit suicide, and I reckon I'll still be somewhere then—and New York seems the best place for me right now."[10]

Caroline's attempt at an independent life was short-lived. About a month after she moved to New York, Allen appeared and they decided to remarry. He then returned to Sewanee to resign and organize the move to New York. Caroline wrote to him, "I cannot tell you how I missed you, but you must know how much I love you. . . . My difficulty was the doubt that you loved me, but that came from a feeling of inferiority I have towards you. It was so wonderful to have you love me that I couldn't believe it was true. But I believe it now, and I want the ring. It will serve to remind us of how important our love is."[11] On April 11, 1946, the Tates were remarried in Princeton in Willard Thorp's study and returned to New York where Allen obtained an editorial job at Henry Holt and Company.

Unfortunately, the Tates' estrangement was not a comedy of errors with the requisite ending in a happy marriage. In the relatively short period from the fall of 1945 when Caroline first left Sewanee until the Tates' remarriage in the spring of 1946, they set the pattern for the remaining fourteen years of their married life. Unable to make a clean break, they were equally unable to live together contentedly for more than a few months at a time. They would separate, often "finally," but during the separations, they would write each other long letters analyzing their marriage and their characters. They would reconcile, but neither their genuine love for each other nor their rational recognition of their weaknesses could overcome the established pattern of their relations. According to Allen, Caroline would become too demanding in her insecurity or violent in her jealousy, while according to Caroline, Allen would be having another one of his "seizures," which sometimes took the form of infidelity. They would again separate, and so the vicious circle continued.

Outwardly, at any rate, the Tates' remarriage appeared to be succeeding during that first year. Caroline was working on her new novel, *The Strange Children* (1951), and was beginning to pick up some prominent assignments, such as her favorable review of Malcolm Cowley's *The Portable Faulkner* for the *New York Times*.[12] Faulkner himself came to dinner since the Tates managed to entertain as usual in spite of postwar shortages. Caroline found Faulkner "exactly the way he ought to be, a little diffident, rather old fashioned in his manners, sort of a touch of Edgar Allan Poe about him. Allen and I both liked him very much. He seems to like Aleck Maury. When he heard that I had written the last Aleck Maury story, in which he foresees his death ["The Presence"], he said 'Don't kill him!'"[13] Apparently Faulkner's courtly charm and admiration for her work caused Caroline to forget the mess he made of her dress while drunk at the Southern Writers' Conference in 1931.

Another pleasing visitor was T. S. Eliot whom Caroline called "literally one of the most charming men I've ever known. He's looking quite ravaged, and no wonder, after what he's gone through, but he's still extremely handsome. And a brilliant conversationalist. He just seems to pluck the right word out of the air, and usually one word would do the trick where somebody else would use a dozen."[14] Eliot's "ravaged" yet still striking appearance after his separation from his first wife sent an ambivalent signal the recently reconciled Tates might well appreciate.

During their time at Perry Street, the Tates also travelled, separately and together. Caroline went to her alma mater, Bethany College, to receive an honorary doctorate. They returned to the Tory Valley for a Fourth of July celebration with some of their old friends, the Cowleys and the Browns. Then Allen went to a writers' conference in Salt Lake City while Caroline received visitors, including her cousin Manny and Eudora Welty. Exhibiting her malicious wit, Caroline wrote, "I do like Eudora so much. I've got to the point where I think she's good looking."[15] The Tates spent the 1946 Christmas holidays in Memphis, visiting the Woods. Percy was in medical school while Nancy tended their son, whom Caroline considered "quite a handsome little fellow."[16] All looked forward to March and the birth of Nancy and Percy's second child, who would be called Allen Tate Wood.

By the spring of 1947, the Tates had separated again and Allen moved into the Hotel Brevoort for a few months. Caroline convinced him to consult a psychiatrist, Dr. Max Wolf, to whom she had dedicated *The Women on the Porch*. According to Caroline, Dr. Wolf found that "an unbalance of the glands—has been responsible for his antics of the past two years" in the sense that this "unbalance" allowed deep-seated problems to emerge. Again according to Caroline, Dr. Wolf found him a "victim of the 'castrative impulse.' He suffered so at the hands of his mother that he goes in deadly fear of women. At the same time he can never fall in love with any woman who couldn't have real power over him."[17]

By the summer, when they were again reconciled, Caroline had decided the blame lay with Allen's job at Henry Holt which prevented him from writing poetry. "When he can't write—and lyric poets can't write all the time—he is in a very devil of a fix and can't keep from taking it out on everybody around him." Although they were together, she tried to keep an emotional distance. "I pay less attention to Allen than I used to. I go ahead and lead my own life, deliberately do things that will give me pleasure, or see people whom I enjoy seeing, even if he doesn't enjoy them."[18] For example, she would spend weekends at Robber Rocks, Susan Jenkins Brown's home in Sherman, Connecticut, where she would delight in gardening and in country atmosphere, but she also participated with Allen in that summer's two-week writers' conference in Salt Lake City.

Caroline also tried to gain some perspective on her marital morass by once again exploring the relations between men and women in a story, "The Petrified Woman," which was published in *Mademoiselle* in September 1947.[19] The setting is based on the Clarksville of Caroline's youth where two young cousins, Sally and Hilda, much like Caroline and her cousin Manny, watch the deterioration of Hilda's father and stepmother's marriage. Cousin Tom uses vulgarity and makes an exhibition of himself when drunk, much to the disgust of his fastidious wife Eleanor. At a carnival, when he sees a woman who was supposedly petrified at the age of sixteen, Tom comments, "I don't know when I've see a prettier woman . . . lies quiet, too." Tom would like a woman to be frozen at the age of romantic idealism, before disillusionment and criticism emerge. For Eleanor it is too late since her cold, judgmental eyes "would freeze" her husband. There is also a suggestion that young Sally is already petrified, too frightened by what she has witnessed to enter the sexual arena herself. Remembering her failure at dancing the previous year, she comments, "I thought it better not to try than to fail."

"The Petrified Woman" might seem to be just another version of one of Caroline's major themes: a man's selfishness petrifies or makes monstrous the women around him. What is different here is that the story actually emphasizes the pathetic effects of a man's conduct on himself, a shift that would continue in Caroline's future fiction. As drunken Tom attempts to rise from the dinner table, he gets his foot tangled in his Cousin Marie's dress and falls to the floor, knocking himself out and cutting himself on his broken wine glass. Although he might like to get away from women, he is inextricably "tangled" with them. At the end of the story we learn that after their divorce Eleanor remarried while Tom burned down his house one night while intoxicated. The picture Sally retains in her memory, the last lines of the story, drive home the message. "Cousin Eleanor, in her long white dress, is walking over to the window, where, on moonlight nights, we used to sit to watch the water glint on the rocks. . . . But Cousin Tom is still lying there on the floor." Eleanor is associated with the revivifying and cleansing properties of water, an increasingly prominent symbol in Caroline's fiction, while Tom is petrified in his self-destruction.

In autobiographical terms, a simplistic interpretation of "The Petrified Woman" would emphasize the "look-where-you'd-be-without-me" message that Caroline might be sending to Allen. While this element is present, she is also performing a deeper self-exploration. In her letters, she often blames her problems with Allen on a pattern of relations with women that he formed with his mother. In "The Petrified Woman," she is examining the effects of her parents' marital difficulties on herself. Like Sally, Caroline was frightened by what she saw and so demanded of Allen constant reassurances to assuage her insecurity. Like Eleanor, Caroline was seeking a way of purifying and renewing her life, and she believed she had found it at Robber Rocks during that summer of 1947.

On November 24, 1947, Caroline Gordon was baptized a Roman Catholic at the Church of St. Francis Xavier on West 16 Street.[20] Her letters and papers prior to her baptism do not indicate that she was contemplating such a step, but her decision could not have been sudden since she would have needed religious instruction before she could be received into the Church. Her silence may simply indicate that she did not want her religious conversion known to her secular, and sometimes cynical, artist friends. In one letter to Ward Dorrance, she provides an account of her conversion.

> I was converted, I suppose, mostly by reading the Gospels. I was reading the Gospel of St. Mark last summer, out at Robber Rocks, and all of a sudden the words that had been in my memory all my life were saying something I'd never heard before. I think I was converted by my own work, too. I have lived most of my life on the evidence of things not seen—what else is writing a novel but that? and my work has progressed slowly and steadily in one direction. At a certain point I found the Church squarely in the path. I couldn't jump over it and wouldn't go around it, so had to go into it. . . . I really would like to tell you what being in the Church is like, but can't. It's like suddenly being

given authority to believe all the things you've surmised. Artists are
fundamentally religious, I suppose, or they're no good.

In an account published five years later, "The Art and Mystery of
Faith," she cites the influences she believed subliminally and gradually brought about her conversion.[21] She remembered a beggar
woman on the streets of Rouen who asked her if she had faith.
"She said it so earnestly and looked at me so intently that I stopped
to consider whether or not I had faith—something I had never
done before in my life. I realized that I had no religious faith of any
kind." She also cited the religious fervor of her tenants, the Norman family, at Benfolly and the example of Dorothy Day. Most of
the piece emphasizes the connection between the artist as Creator
and God as Creator, which in this account is formulated from what
she learned after her conversion from the works of the Catholic
theologian Jacques Maritain.

What all these explanations and influences have in common is
that an outcast of some sort finds an authority, Christianity, that
allows him or her once more to live and work within a meaningful
context. Like the beggar woman, the socially marginal Normans,
and the divorced and drifting young Dorothy Day, Caroline had
recently become a spiritually displaced person: she had lost the two
contexts that gave her life and work its meaning: the belief in a
viable Southern tradition she had mourned and buried in *The Forest of the South* and the person most important to her and most
closely associated with that tradition, Allen. This is not to reduce
her conversion to some sort of consolation prize. A comment Flannery O'Connor made about one of Caroline's later novels, *The
Malefactors* (1956), is illuminating here. Sally Fitzgerald, a mutual
friend, had protested to O'Connor that because two major women
characters in *The Malefactors* join the Church after losing their
men, "the Faith is unintentionally made to seem like chiefly a
refuge for the losers in the battle of the sexes." O'Connor wrote,
"perhaps she [Sally Fitzgerald] is right. On the other hand, some
kind of loss is usually necessary to turn the mind toward faith. If
you're satisfied with what you've got, you're hardly going to look
for anything better."[22]

The effects of her conversion were immediately evident in her
next story, "The Presence," which appeared in *Harper's Bazaar* in

October 1948.[23] The seventy-five-year-old Aleck Maury is now too infirm to hunt or fish and believes "my life's hardly worth living." He identifies with and envies Jim Mowbray, the husband of his boardinghouse keeper, who has just returned from a successful quail hunt. Aleck Maury, however, depends on Jenny Mowbray to nurse him and keep him interested in life, so his loyalties are divided when he learns that Jim Mowbray is leaving Jenny for a young divorcee. Significantly, he compares the grieving Jenny to "a shot bird."

As in "The Petrified Woman," Gordon again emphasizes the damaging effects of masculine roving on the man. Jim Mowbray appears to get off scot free, but Aleck Maury, who spent his life following his pleasure, sport, is now frightened because he can no longer roam and he is losing his refuge since Jenny will close the boardinghouse. Unlike Tom in "The Petrified Woman," however, Aleck Maury is not a figure of despair at the end of the story. He remembers the woman who raised him, his devoutly Catholic Aunt Vic, because, like Jenny Mowbray, her figure was "Junoesque." At the moment of her death, Aunt Vic had seemed to see a presence that was invisible to young Aleck and "he had wondered what it was he could not see." Caroline does not suggest Aleck Maury is converted but merely raises the possibility as he confronts his mortality in the last lines of the story, "Holy Mary, Mother of God, pray for us sinners, now and at the *hour*. . . of our death."

In this ending, Gordon writes the plot that will emerge in much of her future fiction: devout women attempt to lead recalcitrant men to salvation through their examples. The association of Jenny and Aunt Vic with Juno and the Virgin Mary is part of this pattern. Aleck Maury remembers that "Zeus had wooed Io, too, and many another mortal. His wife, Juno . . . suffered from jealousy for all that she were Queen of Heaven." Jenny and Juno do not have the benefits of a Christian faith and both suffer terribly. Aunt Vic is victorious over her sufferings because she has faith in Mary, Juno's successor as Queen of Heaven. Catholics pray to Mary to intercede for them with Christ, and it is as intercessors, like Aunt Vic, that many of Caroline's future heroines appear. Juno herself would reappear with the Greek version of her name in Caroline's final novel, *The Glory of Hera* (1972).

Caroline had not yet succeeded in leading Allen into the Church, but they were engaged together in another project that involved expounding and proselytizing a different kind of doctrine, that of the New Criticism. They were working on an anthology of nineteenth- and twentieth-century short stories, *The House of Fiction* (1950), which would become a staple in the college classrooms of America. The title comes from Henry James's remark that the house of fiction has many windows, many points of view, determined by "the need of the individual vision and by the pressure of the individual will."[24] The thirty "windows" the Tates selected consist of what they considered the masterpieces of American, French, British, German, and Russian writers. Half the writers were American, and almost half of those Americans were Southern friends of the Tates such as William Faulkner, Katherine Anne Porter, Robert Penn Warren, Andrew Lytle, Eudora Welty, and Peter Taylor.

The writers by whose standard the others are judged are Gustave Flaubert, Henry James, and James Joyce, prominent gods in the New Critical pantheon. As stated in the Preface, the Tates believed that the short story had "obtained something of the self-contained objectivity of certain forms of poetry."[25] This is an expression of the New Critical tenet that a work of art is a unified, organic whole that can be analyzed and understood from internal evidence, with little or no assistance from biographical or historical information. Stories that fit this criterion, or aspire to it, are those that the Tates selected.

The popularity of *The House of Fiction* as a textbook was due not only to the fact that the stories were chosen according to the reigning theory of the day, but because for half the stories the Tates provided commentaries in which they explained the techniques by which these writers achieved their effects, particularly the use of point of view. As they state in the Preface, "we assumed that people cannot be taught to write either masterpieces or family letters, but that young persons (of whatever age) can be taught moderately well how to read."[26] Caroline wrote many of these commentaries, some of which appeared first as articles in journals such as *The Hudson Review* and *The Sewanee Review*.[27] Ironically, her fiction was and is relatively unknown, but her standards for fiction helped form the taste of a generation of Americans through *The House of Fiction,* although it contains no story of hers.

Much of the work on *The House of Fiction* was done while the Tates were once again on the move. As they had in the past, Allen and Caroline increasingly turned to changes of location as a relief from marital and artistic tensions. Allen wrote next to nothing while employed at Henry Holt and resigned after two years. Writers' conferences could provide both income and diversion, and so the Tates headed west in the summer of 1948.

They attended a conference at the University of Kansas that also included Malcolm Cowley, Arthur Mizener, J. F. Powers, and Katherine Anne Porter. In attendance was W. J. Stuckey, later a colleague of Caroline's at Purdue. He remembers Caroline as very much a participant in her own right, not merely an appendage of Allen. For example, Arthur Mizener considered her story "Old Red" worthy of analysis before the conferees, and Stuckey remembers Caroline regally attired "for the occasion in an elaborate dress with an amethyst cross around her neck." When Caroline criticized W. Somerset Maugham's story "Rain" at another session, it was defended by a young man in the audience. Stuckey remembers, "Caroline answered him very forcefully, insisting on her view and dismissing his as beside the point." She seemed to Stuckey "a strong woman, very sure of herself, and used to telling anyone and everyone what was what, including and especially Allen."[28]

Caroline's confidence soon received a blow: in Kansas she suffered a hemorrhage and a gynecologist advised her to have an operation to determine if there were a malignancy. Caroline's fright was magnified by her memory of her mother's suffering from fatal breast cancer. Allen persuaded her to have the operation in Ohio in a Catholic hospital, near where he was teaching at the Kenyon School of English at Kenyon College, in Gambier, now the home of his old friend John Crowe Ransom and *The Kenyon Review.* The first operation proved that she had a nonmalignant fibroid tumor, but she had to return to the hospital for a hysterectomy. While Allen was teaching in the School of English, Caroline spent the time recuperating and enjoying Kenyon's "relaxed and informal" ambience.[29] The Tates would return to the School of English for the next two summers, and then follow the School to

Bloomington, Indiana, when it moved there permanently in the summer of 1951.

Back in New York in the fall of 1949, the Tates were planning another sojourn. Allen had agreed to teach in the Humanities Program at the University of Chicago in the spring of 1950. His salary would be sufficient for Caroline to give up her teaching at Columbia that semester and accompany him. The Tates were even able to buy a new car, a Dodge coupe. Before they left, they had a party at their Perry Street apartment in December. In a letter to her daughter, Caroline described that meeting of some of Nancy's "courtesy uncles."

> Uncle Tom Eliot and Uncle John Ransom met here the other night for the first time. They didn't get to talk together much, though, for Cal [Robert Lowell], who came in tight and was pretty trying, sat at Eliot's feet, literally, all evening, blocking off anybody else who wanted to exchange a word with him. Leaving the church or divorcing Jean, or both, has (have) not improved Cal. He is very belligerent, even when he is not tight, and has got awfully worldly it seems to me. He is always ticking off lists of people, poets, chiefly, saying who ranks whom and so on. I imagine the poor boy is rather in a bad way.[30]

Caroline was right, and Lowell was entering a manic phase that would land him in the Payne Whitney Clinic a few months later, but not before he had thoroughly frightened Allen and Caroline.

The Tates, not realizing the seriousness of Lowell's condition, invited him to visit them in Chicago. He arrived on March 30 in what Caroline described as "a dangerous state of elation." Understandably, the Tates became apprehensive about the week he was to spend at their apartment. Caroline wrote, "I have been here alone with him during the day time a great deal and pretty scared at times. I think he kissed me every five minutes (literally.) Towards the last he took to kissing Allen, too, and coming up behind him and squeezing and lifting him off the floor, which scared the hell out of Allen as Cal is very powerfully built and twice Allen's size."[31]

Lowell's nightmarish visit also spawned some unsubstantiated legends. In one, Lowell dangled Allen out the apartment window and made him listen to a recitation of "Ode to the Confederate Dead." In another, Lowell supposedly named all of Allen's lovers to Caroline in an attempt to make Allen reform. From both truth and legend, it is clear that the antic Lowell was attempting to overcome his poetic mentor and assert himself, thus providing a ludicrous living enactment of Harold Bloom's theory of the anxiety of influence.

The visit ended with Lowell's arrest after a "violent scene" in the Tates' apartment that caused the neighbors to call the police. He was released into Allen's custody and departed the next day for a visit to Peter Taylor in Bloomington, Indiana. The Tates warned Taylor by telephone and Lowell was escorted back east by his mother and a psychiatrist.[32] Caroline commented, "It has all ended quite mercifully when you consider what might have happened. . . . Anybody coming in here would think it was we who had been taken into custody by the police. We both walk as if we had bad attacks of rheumatism. Our muscles are still tied in knots."[33]

Aside from this tragicomic encounter, the Tates enjoyed their semester in Chicago. Their apartment at 5631 Kenwood Avenue was sunny, roomy, and "Equipped with every gadget known to Western Man."[34] They finished *The House of Fiction,* and Caroline continued to make progress on *The Strange Children.*

This productive period was followed by their usual summer migrations. In June, they visited Ward Dorrance at his home in Columbia, Missouri. From there they proceeded to a writers' conference in Kansas, and then on to the School of English at Kenyon. In Gambier, they were joined by Nancy and her two small sons, while Percy studied for exams in Memphis. Caroline kept moving in pursuit of lively P-III and Little Allen so that Nancy could take Yvor Winters' course.

Perhaps the Tates could endure, and even enjoy, all these moves because they had provided themselves with a home to which they could return in the fall. That spring they had purchased a house with charming brackets on what was then the outskirts of Princeton, 465 Nassau Street. Since they were again assisted by Allen's prosperous and generous older brother Ben, they named it Benbrackets. Caroline loved it because the "upper storey over hangs the lower storey which gives it that old world, fairy tale look."[35] Although the rooms were small, she felt compensated by the setting, with a terrace, large trees, and a relatively rural neighborhood in which to walk her dachshunds. Princeton itself was in easy commuting distance to New York by train so the Tates were able to keep their part-time teaching jobs in New York, Allen at NYU and Caroline at Columbia.

That fall Caroline busied herself setting the house to rights. She decided to celebrate by combining a housewarming with Allen's fiftieth birthday party on November 19. Eileen Simpson, then married to John Berryman, remembered the occasion. "How they fitted so many people into their doll-sized living room and gave them all food and drink was a mystery. There were such copious amounts of the latter that midway through the evening I noticed that the laurel wreath Caroline had plaited and planted on Allen's head had slipped so that it partially obscured his right eye, externalizing the slippage all of us were feeling at the time."[36] The Tates enjoyed being back with old and new friends in Princeton and remained there throughout 1950, except for a summer round of conferences and visits that included Indiana, Tennessee, Maryland, and Vermont.

All was not gregariousness, however. Both Tates continued to commute to New York to teach. Allen was also at work on what would prove to be an abortive book on Poe. Caroline's project was more successful; by early 1951, she had completed her seventh novel which was published that fall.

In characters and setting, *The Strange Children* is her most closely autobiographical novel. Caroline, Allen, and Nancy reappear as Stephen, Sarah, and Lucy Lewis, the son-in-law, daughter, and grandchild of Aleck Maury. Stephen is in a poetic drought and is writing about the Civil War. Sarah paints a little and tries to manage the servants and house guests at Benfolly. The family receives an unexpected visit from their old friends Kevin and Isabel Reardon, and a poet, Tubby, with whom Isabel is having an affair. The germ of this incident is the unanticipated arrival of Edmund Wilson, Louise Bogan, and her husband at Benfolly in 1931 on an afternoon when the cook got drunk. Kevin and Tubby are composites of many of Nancy's courtesy uncles, including Wilson, Ford Madox Ford, Hart Crane, John Peale Bishop, Andrew Lytle, Robert Penn Warren, and Malcolm Cowley. Isabel is a similar composite of Bogan, Laura Riding, Katherine Anne Por-

ter, Zelda Fitzgerald, Jean Stafford, and others. Numerous anecdotes and events are borrowed from the Tates' lives and those of
their friends.

The intensely autobiographical nature of this novel indicates
that Caroline was using it to judge the Tates' earlier life from the
perspective of her conversion. The verdict was not favorable, as
the title, taken from Psalm 144, indicates: "Rid me, and deliver me
from the hand of strange children, whose mouth speaketh vanity."
The adults in the novel are the strange children; they are so taken
up with the vanities of the flesh and the intellect that they cannot
confront reality directly. As little Lucy notices, "Her father's and
mother's friends hardly ever said what a thing was, they said what
it was like."[37] This is also the way they discuss religion, feeding it
into various intellectual or psychological systems to keep it at a
safe distance from themselves. Sarah comments to her Uncle Fill,
"Of course, everybody isn't as intelligent as you or me, Uncle Fill,
but nevertheless I think it's a good thing for people to have a
religion. I don't think people get along very well without any religion at all" (249).

Sarah's comments, Kevin Reardon's recent conversion, and the
activities of the "Holy Roller" tenant farmers point the way for
Stephen Lewis, who resists it throughout the novel. He enters into
theological discussions with the Holy Rollers and Sarah's Uncle
Fill as intellectual workouts. He says of Kevin Reardon's conversion, "I respect his courage . . . but I don't understand what has
happened to him" (227). He goes to watch the fervent fundamental-
ists handle poisonous snakes as an act of faith and rushes to call the
police and the doctor when one of them is bitten, only to find that
Tubby has run off with Isabel. The last lines of the novel indicate
that Stephen knows he must change his life but is resisting the pain
and difficulty. He thinks of Tubby as "standing at the edge of a
desert that he must cross. . . . He saw those days, those years had
been moving toward this moment and he wondered what moment
was being prepared for him and for his wife and his child, and he
groaned, so loud that the woman and the child stared at him,
wondering too."

The Strange Children is a fascinating novel in that its greatest
strength is also its greatest weakness. The central intelligence is
that of the child Lucy and the use of what Caroline called her

"innocent eye" is masterful in pointing up the adults' futile lives. Caroline named her for St. Lucy, a third-century martyr who is usually depicted as carrying her eyes on a platter but can see more clearly than others; this saint was also invoked by Allen in his poem, "The Buried Lake" (1953). She explained, "I use Lucy's eye more, more than her mind. In fact, I use that hardly at all." The fact that Lucy cannot interpret events makes the novel difficult for the reader to interpret as well. In particular the significance of the shift to Stephen Lewis's point of view at the end of the novel can seem puzzling.[38]

What Caroline said about her life before her conversion also describes the reader's experience of this novel. "In life, as well as in the writing of novels, faith is the key to the puzzle; the puzzle doesn't make any sense unless you have the key."[39] The reader does not need literal faith to understand *The Strange Children,* just a more overt presentation of the need for faith as the key. Caroline herself realized this when she later wrote of this novel, referring to the anecdote in which a black preacher relates his method of composing sermons: "I told them, but I didn't tell them that I was going to tell them or that I had done told them."[40]

In another curious symbiosis of life and art, Allen decided to enter the Church and was baptized at Saint Mary's Priory in Morristown, New Jersey, on December 22, 1950. Jacques and Raissa Maritain, Princeton friends, served as his godparents. Nancy and Percy were present at the ceremony, and they, too, would soon follow Caroline and Allen into the Church.

Caroline was overjoyed, but she still could poke fun at Allen's intellectual approach to Catholicism. She wrote to Malcolm Cowley, "he is having a wonderful time being Catholic and says he doesn't see why he was such a fool as to wait this long to join the Church. Cardinal Spellman has taken the place in his life [as adversarial reactionary] that his old English prof, Dr. Mims, once occupied."[41] Caroline was more interested in rituals and practices than Allen, though both read theology and devotional works. Malcolm Cowley's comment draws the best distinction, though: "I used to say that Allen was a Christian but not a Catholic and Caroline was a Catholic but not a Christian. Of course . . . she had the most malicious tongue of anybody I ever met."[42]

Questions of doctrine were also of great importance to Caroline

in the classroom, but in this setting she was proselytizing the doc-
trines of her craft. Danforth Ross, who took one of her courses in
the late 1940s, remembered her approach: "She started off the
class at Columbia by telling us that the writing of fiction could be
divided into two parts, a part that could not be taught and a part
that could be. 'There is a mystery to the writing of fiction,' she
said. 'There is an irreducible something that you can't put your
finger on.' "[43] For Caroline, what could be taught was technique,
and the student learned it by reading, analyzing, and emulating the
masters of the craft such as Flaubert, Tolstoy, Dostoevsky, Chekhov, James, Ford, and Joyce. In particular, she stressed the rendering of concrete physical details and the vital importance of
point of view, both of which are quite evident in her own fiction.

Often sitting on the front of her desk, she taught technique
through the example of her own close readings of illustrative texts.
A favorite was the seduction scene in *Madame Bovary,* which her
students all seem to recall as a masterful performance. Robert
Kettler, who took a course from her at Purdue in 1963, described
her method.

> In class she worked closely with a text, using it as an artifact, not to
> explicate or interpret but to examine as an example, a way of doing it.
> Seeing how others had solved problems. She did not lecture, she read
> and analyzed, whether it was student work or something by one of the
> masters. Her comments were always pointed and explicit. None of this
> vague 'I don't think this really works here.' She said why.[44]

A student at Columbia in 1947, Jean Detre, said that "essentially
she taught the class how to read."[45]

Danforth Ross, among others, found her "rather dogmatic in the
classroom. . . . she is very sure of her direction and the minute a
student starts eating grass over at the side of the road she just
about jerks his neck off. 'I'm not going to waste my breath discussing that until you show some glimmer of getting this,' she once said
to a student who showed reluctance to go down the road with
her."[46] This fervent desire to teach her craft, to put art first, could
make her seem cold and insensitive to her students' feelings, as
Jean Detre recollected.[47] Robert Kettler remembered his own pillo-
rying in the name of the muse. After much labor, he submitted a
twenty-page story. "The next class she pulled it out and said, 'Now

this story has *one* good paragraph in it that I want to read to you; it has good sentence variety.' " She was unable to find the paragraph and "went on to some other text to illustrate her point about sentence variety. No apparent sense of how mortified I was or how crushed by her casual dismissal of my efforts. She was concerned with the making of fiction, and our feelings were not really an issue."[48]

Although this might be a bit much to ask of adolescents and young people, those who could put their egos aside found her a superlative teacher. Despite his experience, Kettler called her a "great teacher," and added, "She taught me to have a reverence for the text . . . and what hard work writing really is. And that writing well, honoring the tradition of those who had gone before you was important."[49] Marjorie Kaplan, who also studied with Caroline at Purdue, believed that her dogmatism also supplied her strength as a teacher: "I attribute the fact that she was a great teacher to what was, to me, a negative characteristic. She seemed to be an elitist—elitist about people's qualities, art, intelligence, the South, even Catholicism. And yet she was democratic in her elitism; she seemed to enroll us, as members of her class, into the elite, so that often we were led to do better work than we were capable of on our own."[50]

Whatever the differing opinions about Caroline's classroom approach, near unanimity reigns about her extreme generosity with her time and commentary in individual conferences and in grading student work. Marjorie Kaplan's experience is typical of many.

> She held individual conferences with each student and would work an hour at a time trying to get us to 'render' what we were trying to say. She could recognize the glimmer of an idea behind a sentence and surprise us with it. She emphasized details, details, details, and, at the same time she would quickly eliminate any glib or dull ones. Then, once again, each of us would trudge back to her office with alterations on this same story and she would reward our work with praise—always recognizing where we had labored—and then go on to the next phase of the story. The smallest kernel of true writing never escaped her.[51]

These intensive conferences would often be supplemented by pages of close analysis laboriously typed out by Caroline; many of these commentaries are kept and treasured by her students to this

day. Obviously, this kind of devotion to teaching cut into her time
for her own writing and is one reason for her diminished productivity in the 1950s and 1960s. Two "pupils" certainly recompensed her
efforts. Flannery O'Connor and Walker Percy could perhaps name
Caroline what Eliot called Pound, "il miglior fabbro," the better
maker, because she helped teach them their craft.

Percy and O'Connor, both Catholics and Southerners, coinciden-
tally sent Caroline some of their work for commentary in the
spring of 1951, when Caroline was in the full flush of her recent
conversion. After a decade-long study of the existentialists, particularly the Christian existentialists Kierkegaard and Marcel, Percy,
in his thirties, was beginning to write his own philosophical essays
and to embody his ideas in fiction. He sent Caroline his first novel,
The Charterhouse, a hefty work of 942 typed pages. She replied
with thirty pages of the kind of comments and suggestions she
lavished on her creative writing pupils.

Although the length of her commentary suggested that the manuscript had its faults, Caroline enthusiastically recognized Percy's
genius. She wrote to Malcolm Cowley, "I think that he is one of
the most important young writers coming along. . . . I prophesy
that he will do great things." With Allen, she had the manuscript
sent to Jack Wheelwright at Scribner's, who succeeded Max Perkins as her editor after Perkins' death. Wheelwright replied that
the manuscript needed extensive revision, and it was never published.[52] After another abortive attempt at a novel, Percy published *The Moviegoer* in 1961, the novel for which he won the
National Book Award, fully vindicating Caroline's judgment. She
continued to follow his career with pleasure, reading each of his
novels as it was published.

A comment Caroline made to Wheelwright about *The Charterhouse* illuminates her attitude toward both Percy and O'Connor:
"The book is an example, I suppose, of the trend towards orthodoxy in religion which so possesses the creative imagination today,
particularly in the young writers. Whenever I read a novel by a
beginning writer today I am astonished by the difference between
his situation—not necessarily his attitude—towards orthodoxy or
religion in general, as opposed to my own situation when I began
writing novels over twenty five years ago."[53] Caroline envied them
the security of their faith at the beginnings of their careers, in

contrast to her almost two decades of writing prior to her conversion. Her envy, however, was of the wistful and constructive type, perhaps promoting in them something of the writer she wanted to be, particularly in the case of Flannery O'Connor.

In the spring of 1951, O'Connor was completing her first novel, *Wise Blood*. Because she suffered from lupus she lived on her mother's farm in Milledgeville, Georgia, remote from the literary world. She gratefully accepted the suggestion of her friend, poet and translator Robert Fitzgerald, that he send the manuscript to Caroline for her advice. Caroline sent O'Connor her commentary, which does not seem to have survived, but Caroline's initial response to her work is available through a letter she wrote to Robert Fitzgerald. "I'm quite excited about it. This girl is a real novelist. (I only wish that I had as firm a grasp on my subject matter when I was her age!). At any rate, she is already a rare phenomenon: a Catholic novelist with a real dramatic sense, one who relies more on her technique than her piety."[54]

Fortunately, Caroline's comments on the next draft of *Wise Blood* do survive. As was typical of her analyses of apprentice work, the criticism is largely confined to matters of technique. She cautions O'Connor about expecting too much effort from her reader, the overuse of landscape to echo a character's feelings, and the artificiality of a character's uttering a speech of more than three sentences. Her praise, in contrast, expresses her own new goal for herself as a Catholic writer, to use the symbol to link the natural with the transcendent. She wrote to O'Connor, "you are like Kafka in providing a firm Naturalistic ground-work for your symbolism. In consequence, symbolic passages—and one of the things I admire about the book is the fact that all the passages are symbolic, like life itself—passages echo in the memory long after one has put the book down, go one exploding as it were, depth on depth."[55] Caroline continued to read and criticize O'Connor's work before publication until O'Connor's death in 1964.

Caroline had need of the steady anchor of faith she found in Percy and O'Connor's works because the Tates' lives were once again in a state of flux. Despite their recent purchase of Benbrackets in Princeton, Allen was lured to the University of Minnesota to replace Robert Penn Warren in the fall of 1951. Prior to the move, Caroline accompanied Allen to the Indiana School of Letters that summer, and then back to Princeton to prepare for the move. The Tates rented a large old house at 1801 University Avenue in Minneapolis from Professor Joseph Warren Beach, who was away for the year. Caroline wrote to Nancy, "I have just discovered that we have an oratory: the little bay off the livingroom. . . . It makes a splendid oratory. I keep my breviary there and say Matins in the morning while my egg is boiling, then say Lauds after I have consumed the egg."[56]

Both Allen and Caroline seemed to like Minnesota a great deal. Allen was pleased with what he perceived as the tranquil state of departmental politics and by February was writing poetry again.[57] In the early 1950s, essentially his last productive period as a poet, he completed "The Maimed Man," "The Swimmers," and "The Buried Lake." Caroline began her new novel, *The Malefactors* (1956), but she varied her labors by teaching a creative writing class for twelve young women at nearby St. Catherine's College. The group included an auditor, Abigail McCarthy, wife of Congressman Eugene McCarthy.[58] Of the prospect of teaching at St. Catherine's, Caroline wrote, "To tell the truth I was pretty damn sick of teaching under Protestant aegises. It makes you heart-sick—or at least it makes me heart-sick at times, to contemplate the amount of creative energy that goes down the drain."[59]

The first year at Minnesota was also cheered by two arrivals. Dorothy Day appeared at the Tates to take a rest, midway through her tour of Catholic Worker houses throughout the country. Caroline was struck by how little clothing she carried and by her box of reliquaries. She wrote to Nancy that Dorothy Day "left Allen a bit of flesh. . . . of Pius X done up in a handsome reliquary to help him finish his poem. He rejected a reliquary of the Little Flower [St. Therese of Lisieux] all done up in roses on the ground that it might cause his style to deteriorate." The other arrival was a grandchild, the third child of Nancy and Percy Wood, who was born on January 31, 1952, and named Caroline for her maternal grandmother.

The Tates' usual complicated summer migrations included Allen's trip to Paris in May where he attended the Congress of Cultural Freedom with Faulkner, Porter, Auden, and other luminaries. From there he proceeded to London for a week. By midsummer he was at the American Academy in Rome and in August had an audience with the Pope.

Caroline was equally on the move, making a round of visits that extended from Utah to the East Coast and that included Ward Dorrance, the Cowleys, and the Woods. She also made a retreat at Dorothy Day's Maryfarm in Newburgh, New York. According to Dorothy Day, Caroline seemed to be sending ambivalent signals about herself: "She wears a dress like a bathing suit, short, backless. And she brings a lovely statue of the Virgin and Child."[60] Of her travels, Caroline wrote to Allen, "I do not want to go through any more summers like this one. I don't think I've stayed in place more than five or six days at a time—and nothing to show for it but fatigue. Well, a little visiting with the children."[61]

This summer of separate travels anticipates the pattern of the rest of their married life. In the fall of 1953 the Tates were back together again in a house they rented in St. Paul at 1908 Selby Avenue. In later letters to each other, they discuss this as a period when they felt estranged from each other, and Caroline believed Allen was rejecting her. She sought consolation in religion, enjoying her predominantly Catholic neighborhood and her proximity to St. Mark's Church.

Once again the Tates seemed to feel more comfortable separated, and for the spring of 1954 Caroline planned a visiting lectureship at the University of Washington at Seattle. Although the lectureship would give them a respite, the Tates planned to try a change of scene again, since Allen was arranging a year in Rome for them under the auspices of the American Academy. While she was in Seattle, he used St. Paul as his home base, but embarked on a series of lectures and readings to get their finances in order before the trip.

Caroline's sojourn in Seattle was a productive one. Her academic duties were light, and she spent much of her time on the roof of her apartment building, writing and sunbathing.[62] In this period, she was attempting to merge her beliefs in Christianity and art in both her criticism and her fiction. For this effort she relied on Jacques Maritain's *Art and Scholasticism,* but Caroline was not fundamentally a theorist.

She derived a standard for judging her own work and that of others that boiled down to Christianity and craft. In a review of two books about Willa Cather, she praised Cather's "heart wide enough to embrace all humanity," but faulted her for being "aston-

ishingly ignorant of her craft."[63] In "Some Readings and Misreadings," which appeared in *The Sewanee Review*, Caroline argued
that "in the nineteenth century and in our own century as well the
fiction writer's imagination often operates within the pattern of
Christian symbolism rather than in the pattern of contemporary
thought. The peculiarly Christian element of the great nineteenth
century novels is their architecture. Many of them are based on the
primal plot: the Christian scheme of Redemption."[64]

In an article written in Seattle, "Mr. Verver, Our National
Hero," which was published in *The Sewanee Review* in 1955, she
applies that thesis to Henry James's *The Golden Bowl*. Perhaps
thinking of her own admiration for her father, Caroline did not
find the relationship between Adam Verver and his daughter Mag-
gie "pathological" because James makes "Mr. Verver so superior
to the average man that his daughter would be a fool not to find
inspiration in him." The Ververs, she argues, save themselves and
their spouses through the exercise of *caritas*, Christian charity or
love. For this reason, Caroline considers *The Golden Bowl* the
zenith of James's art, "the only one of his major creations that is a
comedy in the sense that Dante's great poem is a comedy, the only
one in which virtue is triumphant over vice."[65]

In her account of her conversion in the *Newman Annual* for
1953, "The Art and Mystery of Faith," Caroline states baldly, "I
have come to believe that the writing of [serious] . . . fiction is in
essence a religious act. We are moved to imitate our Creator, to do
as he did, and create a world. We are, of course, attempting to do
the impossible when we do that."[66] In "Emmanuele! Emmanuele!," the story she wrote in Seattle during the spring of 1954 and
published a year later in the *Sewanee Review*, she explores the
relationship between Christianity and craft. She also illustrates her
point that, unlike God, the artist cannot really create a world from
nothing, but merely constructs an imaginary world from what God
has provided. Caroline obtained the germ of the story through
Walker Percy's suggestion that she read Andre Gide's correspondence with Paul Claudel, which, in turn, led to her reading of
Gide's memoir of his wife. Her story's Guillaume Fay is loosely
based on Gide, and her Raoul Pleyol was suggested by Claudel.[67]

The central consciousness of "Emmanuele! Emmanuele!" is not
that of Fay, however, but Robert Heyward, a somewhat naive

associate professor of English who "wrote poetry 'on the side.' "[68] Through an old friend, he obtained the position of Fay's private secretary in North Africa for the remainder of his sabbatical. Heyward is in awe of Fay as a consummate artist who lives for his craft. In particular, he admires the way Fay, for the past thirty years, has written to his wife each day at noon, whenever they have been separated. Heyward wants to emulate Fay and tries to write every noon to his own wife in America.

Heyward is so blinded by his artist worship that he does not heed various warnings about Fay's character. Mme. Rensslaer, a widow who is Fay's cousin, confides to Heyward that she refused to marry Fay in her youth on the grounds that a man who wrote every day in his journal while staring into a mirror was not a good prospect. Instead, she married M. Rensslaer and was quite happy with him until he died. Another old and former friend, the diplomat-poet Raoul Pleyol, also warns Heyward about Fay. He perceives Fay's self-examination as a sinful narcissism because "An artist's first duty is the same as any other man's—to serve, praise, and worship God" (337). These admonitions are reinforced by Fay's association with beasts and snakes in Heyward's subconscious mind.

On their return to Fay's country chateau in France, Heyward finally meets Mme. Fay. She turns out to be a worn looking woman who appears much Fay's elder, rather than the several years senior she actually is. Her appearance has been sacrificed to nurturing her garden and various needy people. Fay brings her some beautiful gloves from North Africa, but instead of using them to cover her weathered hands she gives them away to a secretary. Her laugh, though, belies her battered exterior; "it might have come from a child who had laughed out in exuberance and might laugh again any minute"(339).

When Fay asks for the letters he has sent his wife over the past three decades, he learns that she has destroyed them, and he emits "a whimpering sound, such as might be made by a dog in distress . . . *or some other beast, that has thought to escape, being forced over a cliff*"(350). The last lines reinforce the fact that Fay's selfishness has damned him; he has lost the angelic component of his nature and only the bestial remains: his eyes were "twin prisons in which a creature that had once sported in the sun would sit forever in darkness."

According to Caroline's account to her daughter Nancy, the wife "found the letters false and burned them. It is the kind of thing only a saintly person could do without [sic.] impunity."[69] As in the case of *The Strange Children,* only a Catholic reading provides the key to what otherwise would seem a puzzling act of philistine cruelty. Mme. Fay had lived up to Fay's name for her, Emmanuele, and tried to free him from his enslavement to himself. An artist who is false to his duty to God and his fellow man is only capable of producing a false art.

This story's suggestion that a religious life is so much more important than art may also help account for Caroline's slower rate of publication over the next decade or so, as did Allen's response to the tale. Caroline was elated on finishing "Emmanuele! Emmanuele!" and sent it to Allen for his comments, which apparently were not enthusiastic. She wrote to him, "I was a little dashed at first by your luke-warm reception of the story, but on second thought was pleased. If it's not as good as 'Old Red' it would seem there's not much use in my writing any more stories. This one nearly killed me. The thought of another one positively paralyzes me."[70] Caroline never did publish another short story; her remaining short fictions were actually excerpts of novels in progress.

Allen's "luke-warm" response and Caroline's extreme reaction to that response may arise from the story's autobiographical aspects, for which the obvious borrowings from Gide's life serve as red herrings. Fay is endlessly attempting to work on a long poem he cannot complete, much like Allen during this period. Mirrors also appear prominently in some of Allen's poems. Caroline, like Mme. Fay, was older than Allen, but increasingly looked even older in contrast with him. Also like Mme. Fay, she was an avid gardener and liked to take in those friends or relations who needed help. The implications of Mme. Fay's choice of life and her destruction of Fay's writing were probably not lost on so astute a critic as Allen Tate.

In his poems, Allen was also evaluating his past, including his life with Caroline. "The Buried Lake" is a long and complex poem that concerns the poet's Danteesque journey with Saint Lucy as his Beatrice. He remembers the hotels of his childhood and his attempts to play the violin. In the music room of one such hotel, he encounters

A stately woman who in sorrow shone.

I rose; she moved; she glided towards the hall;
I took her hand but then would set her free.
'My love,' I said.—I'm back to give you all,'

She said, 'my love.' (Under the dogwood tree
In bloom, where I had held her first beneath
The coiled black hair, she turned and smiled at me.)

I hid the blade within the melic sheath
And tossed her head—but it was not her head:
Another's searching skull whose drying teeth
Crumbled me all night long and I was dead.[71]

The "coiled black hair," the stateliness, the way her eyes always shone in pictures as if she were sorrowful or in tears, as well as the reference to young love, all point to this figure as a version of Caroline. As in the months leading to the Tates' first marriage, their response to each other is ambivalent: she moves away, he holds her, lets go, but then finally commits himself by promising her endless reassurance, "I give you all." Despite their mutual pledges of "My love," they destroy each other. He finds that she is a Medusa, with "coiled black hair" and cuts off her head. She is *la belle dame sans merci* who keeps him to his promise of giving her all by turning into a death's head who consumes him with her "drying teeth," the voracious need for him that he feels dried up the source of his poetry.

As in the instance of the autobiographical aspects of "Emmanuele! Emmanuele!," this small passage from "The Buried Lake" is a mere fragment of a rich and complex work. In terms of the Tates' marriage, however, both works are signals of despair. In her story, Caroline is suggesting that Allen is killing himself and his art; in "The Buried Lake," he implies her devouring personality destroyed him.

These distress signals through art may have been necessssary because the Tates remained separated for much of the summer of 1954. Allen attended some writers conferences and then proceeded to Oxford for a symposium on American culture, where, following her own round of conference and visits, Caroline joined him at the end of July. After a brief stop in Paris, the Tates arrived in Rome at the beginning of September and were installed in an apartment in the Villa Aurelia, complete with a devoted maid named Assunta.

Allen's first lecture was a success, and the Tates began a social round with old and new friends. The American Academy was a magnet for visiting American artists. At one party that fall, the Tates saw three novelists: William Faulkner, Elizabeth Spencer, and Anya Seton. They developed a close friendship with Franklin and Ida Watkins, the painter-in-residence and his wife. Not all was congeniality, though; Allen found himself and his new Austin mistakenly mobbed in an anti-British demonstration, but escaped with the help of some friendly passers-by.[72]

The Tates, particularly Caroline, delighted in seeing the sights. For Caroline, Rome was, of course, the seat of her newfound faith. As she put it, "Roma is my homa."[73] The Tates did venture outside of Rome as well, visiting many of the principal Italian cities, including Fiesole, where Robert and Sally Fitzgerald were living with their children. Venice for Caroline was "too much of a confection . . . but the Tuscan landscape is the country of my dreams—the country you always knew existed."[74]

She also revelled in art: "The mosaics in Ravenna had me reeling."[75] The noted art critic Bernard Berenson invited them to visit him at his villa in Florence, which Caroline found overwhelmingly lavish and his art gallery "fabulous." Her novelist's eye was most intrigued by the aged owner himself: "Berenson is very small, bearded and moves with a kind of grave dignity in the aura with which he has surrounded himself. . . . [He] still has all his faculties, but inside that aura of which I spoke he is furious and dismayed. He is afraid, I think, that he may die without finding out what it is all about."[76] Once again, from her new religious perspective, Caroline was judging someone whom she believed made a religion of art.

Underneath this surface of sociability and sightseeing, the Tates again found their relationship deteriorating. Caroline believed that since Allen was never able to write while abroad, he was drinking too much. She responded with anger, which started the process again. In an attempt to break out of this cycle, she obtained the name of a Jungian analyst, the Dottoressa Dora Bernhardt, and began to see her regularly. Caroline believed Dr. Bernhardt had "one of the most powerful and subtlest imaginations I ever had to do with."[77] Allen, infected by Caroline's enthusiasm, soon initiated sessions of his own.

According to Caroline's accounts, Dr. Bernhardt stressed the analysis of dreams as guideposts from the unconscious mind that pointed the way to the healing of the psyche and a further stage of development. Caroline recorded her dreams on her typewriter, as she did intermittently over the next few years. Basically, the Tates relearned in Jungian terms what they already knew in Freudian terms and in the language of their own art. Allen's fear of committing himself to Caroline derived from the smothering relationship he had with his mother; because he was devoting his psychic energy to fending off what he perceived as Caroline's attacks, he was often unable to write poetry. Caroline's need for endless reassurance stemmed from her ambivalent feelings toward her own mother which she projected on her marriage: she believed her mother rejected her because she resembled her father, yet she sympathized with her mother's inability to achieve a stable existence because of her father's roving proclivities.

Once again, no amount of intellectual understanding seemed to allow the Tates permanently to escape their mutually reinforcing, destructive patterns of behavior, though they did manage to spend the next academic year together. What Caroline did gain from her analysis was the germ of her last novel, the unfinished and unpublished, *A Narrow Heart: The Portrait of a Woman*. Because of the Jungian emphasis on the mythic patterns that arise out of the collective unconscious, Caroline wanted to evaluate her life and that of her ancestors in mythic terms. Her last published novel, *The Glory of Hera* (1972), is an exposition of the myth of Hera and Heracles, which was to provide a context for her "autobiography," *A Narrow Heart*.

Her analysis was cut short, however, in the spring of 1955 when she learned that Nancy was seriously ill. She returned immediately to Princeton to take charge of the Wood household, still quartered at Benbrackets. Allen decided to remain behind because he felt that he should not lose his May and summer salary in case Nancy's illness necessitated financial aid as well. Again, the Tates seemed to thrive apart. Aside from his anxiety over Nancy, Allen was happy in his analysis and his life at the Academy. Caroline seemed a virtual powerhouse of energy and initiative. She organized the household and the care of the three children, and even started driving again after a hiatus of some years. More important, she believed she needed to take action to get to the root of the problem.[78]

Since she believed Nancy's illness was at least partially induced by the strain and work of raising her family in Benbrackets' cramped accommodations, she decided that the Tates and the Woods would find a house in Princeton large enough for all of them. Nancy, who was recuperating well, was able to join her in house-hunting, and they found what they considered the ideal house at 54 Hodge Road. Caroline referred to it alternately as "the Hodge Horror" or "Dulce Domum," the latter being the inscription on the mantel. The horror of 54 Hodge Road was its late Victorian architectural exuberance and its condition: much paint and renovation were required and the garden had reverted to nature. Its sweetness, however, was the offbeat charm of that same architecture and its size: six bedrooms and three and a half baths, leaving plenty of room for an upstairs suite with its own kitchen for Caroline and Allen. The move was made and the painting and repairs commenced in May 1955.[79]

As in the case of Benfolly and Benbrackets, Caroline put much labor into a house in which she would live very seldom, though in this case, she had her satisfaction in the Woods' happiness in their new home. The Tates returned to Minneapolis in the fall where they stayed with another faculty couple until the house they wanted to rent was available. The house at 1409 River Road, near the Franklin Street Bridge, was worth waiting for because it faced a boulevard beyond which was the Mississippi River. The Tates, as ever, enjoyed seeing old and new friends connected with the University of Minnesota: the Danforth Rosses, the Thomas Mabrys, the Leonard Ungers, Sam Monk, and John Berryman.

During that academic year of 1954–1955, both Tates were reading Jung while pursuing their own writing. Allen was at work on a poetry anthology, interrupted by a three-week lecture tour on the West Coast in March. Caroline was in the final stages of her novel. By the summer of 1955, at the School of Letters in Bloomington, Indiana, she believed she had completed it, but gave it further revision in August while babysitting for the Woods in Princeton.

In many ways, *The Malefactors* is a reworking of the themes of

"The Petrified Woman," *The Strange Children,* and "Emmanuele! Emmanuele!" with supernatural and Jungian overtones. The central consciousness is that of Tom Claiborne, a poet who has written only eight pages in ten years. Although the reader is decoyed by Claiborne's reference to Allen Tate as a poet he used to know, Claiborne is clearly based on Allen in a number of ways, including a friend of his youth, Horne Watts, obviously modelled on Hart Crane. Claiborne's family history, however, is that of the Meriwethers, suggesting that through Claiborne's journey to the Church, Caroline is once again relinquishing her family and the South's burden of the past; all the Claiborne land is now under water as part of a government water program in the same way that Claiborne's dead past will now be reborn in the waters of faith.

Claiborne's wife, Vera, like Mme. Fay of "Emmanuele! Emmanuele!," resembles Caroline in her circle of lame ducks and her love of the land. The relationship between Claiborne and Vera also has some similarities to that of the Tates, especially Allen's retreats before Caroline's need for reassurance. Claiborne remembers that Vera "had come to him as to a physician—or a magician. He was to give her everything she had ever wanted and never got."[80] Tom Claiborne, like the Tom in "The Petrified Woman," wants his wife paralyzed, docile and undemanding: "If only she had been content to remain what she was when he had found her, a bird fluttering on a terrace that a man might pick up and warm in his bosom, a bird that would nestle tamely, grateful for any warmth it might come by, and not be always turning its fierce golden eye on yours, not always be beating its maimed wing against your breast" (190).

Vera is more the woman Caroline would like to be than a version of Caroline as she was. Vera wanted to destroy the late paintings of her artist father, in which he posed naked before a mirror, as Mme. Fay, in what Caroline considered a saintly gesture, had destroyed Fay's narcissistic letters. Vera, as her name suggests, represents the path to true art, Christian art; she will serve as Beatrice for her clay-born husband's journey to faith.

Claiborne is aware of the futility of existence on their farm, "Everybody busy all day long doing nothing" (7). He keeps hearing a Voice, perhaps supernatural or perhaps the voice of his conscience or better self, which warns him that he is destroying himself, but he wants to suppress it. The fact that Claiborne is at a fork in his life's journey is represented by two women who reappear in his life after years of absence. Catherine Pollard, based closely on Dorothy Day, visits the farm; she is no longer the libertine Claiborne had known in her youth and he is drawn to her evident saintliness. The attraction of another woman, however, is temporarily stronger. His wife's younger cousin Cynthia comes for a visit and seduces him; perhaps as Caroline perceived Allen seduced by her younger cousin in 1933. As her name suggests, she is fickle and false reflected light, like a mirror or the moon. She attracts him because she does not reflect his true self, but merely flatters his false self as influential man of letters.

When he finally realizes Cynthia's falseness and fears that he is falling apart, Claiborne goes to seek Vera at a place that resembles Dorothy Day's Maryfarm on the Hudson River. He speaks to her, but cannot understand her conversion or her new way of life. As he leaves, an alcoholic priest reminds him that Vera must listen to him because a wife is subject to her husband as the Church is subject to Christ. When he returns to the Chapel of St. Eustace, Catherine Pollard's New York City base, she tells him the same thing. The epigraph for the novel, a line from Jacques Maritain, expresses a similar sentiment: "It is for Adam to interpret the voices that Eve hears."

These statements are calls for obedience to just authority, which Claiborne cannot claim until he realizes that he is clay born, a malefactor, weakened by original sin like everyman. When his reason is strengthened and revivified by faith, it will be able to interpret correctly the imaginative and intuitive messages of his wife or his muse. The last line of the novel suggests that Claiborne is indeed on the road to his rightful place as a Christian artist when he pictures his return to Maryfarm: "He could be sitting there on the bench with the other bums when she came down in the morning."

The Malefactors indicates Caroline's increasing ambivalence toward her role as an artist who is also a woman. If Adam has the interpretive role, why is Eve writing all these novels? Caroline was conscious of this dilemma, as she wrote in a letter to Ward Dorrance about the time she was completing *The Malefactors*.

. . . while I am a woman I am also a freak. The work I do is not suitable for a woman. It is unsexing. I speak with real conviction here. I don't write 'the womanly' novel. I write the same kind of novel a man would write, only it is ten times harder for me to write it than it would be for a man who had the same degree of talent. Dr. Johnson was right: a woman at intellectual labour is always a dog walking on its hind legs. When you add to that the task of running a house, serving dinner that seems to have been prepared by an excellent cook, and all the while trying to be a good hostess—which means trying to make every man in the room have a good time—oh well, I am inclined to self-pity now and I don't deserve any pity at all, for I have a good time in this life. But I do have a lot on my hands. I bite off more than I can chew all the time.

The pity here is that this intelligent, spirited, even feisty woman ultimately felt she had to accept the verdict of her male-dominated culture and religion and regard herself as a freak or a dog on its hind legs. Nearly sixty years of age, she appeared to be running out of energy to fight those cultural values to follow the demands of her artistic imagination. She wrote to Allen on November 6, 1955, "My work as a novelist is over." *The Malefactors* was the last piece of imaginative fiction she ever wrote. *The Glory of Hera* is a retelling of myth and the fragments of *A Narrow Heart* are autobiography or family history.

"It is for Adam to interpret the voices that Eve hears." This is an epitaph for her career as an artist, and she chose it as the epitaph for her life. It is graven on her tombstone.

Despite the fact that *The Malefactors* is a statement of submission to patriarchal values, some of the living models for the characters did not perceive it as such. Dorothy Day was shown the manuscript by her and Caroline's mutual friend, Susan Jenkins Brown. Dorothy was horrified by the references to Black Masses and alchemical experiments using consecrated wine in Catherine Pollard's wild youth since they were not only untrue but blasphemous. She wrote to Caroline's editor at Harcourt Brace and succeeded in having the reference to the Black Mass deleted. She also did not want the novel dedicated to her as Caroline had planned.[81] Although one might argue that *The Malefactors* is a work of fiction and must be judged as such, Caroline is so close to fact in the character of Catherine Pollard that Dorothy Day's perturbation is certainly understandable.

Adam-Claiborne-Allen was also somewhat upset. After all, the last line of the novel has him characterizing himself as a "bum" without his wife. To his credit, Allen did not wrap himself in the doctrine that a woman should be subject to her husband, but the somewhat stinging portrayal rankled, particularly after the novel was published and was often treated as a literary guessing game, a *roman à clef.*

Unaware of her novel's future repercussions, Caroline handed it to her editor in September 1955 and returned to Rome for more analysis with Dr. Bernhardt. Allen had spent the summer in Rome doing just that, but the Tates did not even meet in passing. They did, however, continue their dialogue of mutual analysis in letters throughout the fall, with as little progress as before toward restoring their marriage. Caroline made plans to teach during the spring semester of 1956 at the University of Kansas without seeing Allen in Minnesota before she went out to Lawrence. She wrote to him, "If I did, it would be falling into that 'trap' the Dottoressa talks about."[82] Perhaps another gain from her analysis is an increasing self-reliance that would cause her to choose to live alone rather than live in conflict.

Self-reliance was not easy, however, and the spring of 1956 was a particularly unhappy time for Caroline. She had a demanding program at the University of Kansas, two classes and a weekly public lecture. Although she would use the lectures she wrote as the basis for *How to Read a Novel* (1957), she found the weekly task of composing and delivering them quite a strain. Her spirits were further depressed by the responses to *The Malefactors.* Caroline resented her novel being read as a *roman à clef* and felt the reviewers were prejudiced against it for its Catholic overtones. She was somewhat heartened by a letter she received from Jacques Maritain in which he praised her use of actual people as adding another dimension to the novel.[83]

Worst of all, her two reunions with Allen that spring, one in Kansas and one in Minnesota, were not successful. She wrote to Allen on April 11, "I am sorry if I seemed unsympathetic and hard while I was in Minneapolis. I know I was. That night you appealed to me I was just stubborn because I was drunk. I am worried to death about you." Her anxiety over Allen was compounded by the sense that she was no longer able to cope. "Darling, I am going through a strange and terrible time, so please be patient with me. A certain enthusiasm, a kind of vigorous response to life which I held on to through all sorts of trials has left me. I can hardly face each day. I long only for the night."[84]

The Tates decided to try again anyway, meeting first at the Fugitive Reunion at Vanderbilt in May. Caroline did not spend all her time in Tennessee at the formal sessions and official parties. She saw Benfolly, which looked as if its new owners had improved it greatly. She visited her cousin Manny, Marion Meriwether, and also spent time at the Cheneys, enjoying their lavish hospitality. "But the most memorable evening," she wrote, "was when John [Ransom], and Don [Davidson] and Allen and Red [Warren] and Merrill Moore read their poems." She concluded about the Reunion, "Oh, Lord it was so much fun."[85]

At the beginning of June, the Tates returned to Princeton where they lived at a house at 145 Mercer Street and tried to decide what to do. From Kansas that spring, Caroline had written to Allen that her life as a visiting professor and "these separations that we have had keep me suspended in a sort of half-life, forcing me to live alone and live in a way I would never choose." Her first priority that summer was to establish a more settled life for herself, to find "the fixed abode" that she realized she craved during her sessions with Dr. Bernhardt.

Caroline found an old house at 145 Ewing Street, Princeton, which seemed ideal to her, as she wrote to Malcolm and Muriel Cowley. "One part is Revolutionary. One part was built in 1830. There are three ancient working fireplaces. The rooms have ideal proportions." What she liked best, however, was the yard. "The house—painted barn-red, stands flush with the street, so that all the land is in the back, nearly an acre. The long stretch of greensward terminates in a little wood, which has two willows towering over it." Despite its proximity to the center of town and a shopping center, "The Red House," gave the impression of being in the country. Caroline planned to "stay here all year and garden, garden garden."[86]

This decision to garden was also a tacit admission that she and Allen would remain separated. By the end of the summer of 1956, Allen moved out of the Red House and into the Princeton Club in New York. He felt he was too old and tired to receive Caroline's suspicious scrutiny, answer her demands for reassurance, and narrow his circle of friends to those who would not remind her of past troubles.[87] Caroline replied, "I think it is quite possible that we may never be able to live together again."

That was indeed the case. After Allen's State-Department-sponsored trip to India that fall, she presented him with a formal agreement for separate maintenance that she had a lawyer compose. Allen consented, and rented an apartment for himself in Minneapolis. Caroline concluded, "I can stop worrying about him actively for a while, which will be the one thing needed to let me get on with my book."[88]

With the assistance of Malcolm Cowley and Susan Jenkins Brown, Caroline was transforming her Kansas lectures into *How to Read a Novel,* which she dedicated to Sue Brown. *How to Read a Novel* is really a written version of the way she taught her students to read fiction. It contains explanations of techniques, such as point of view and tone, and examples from Caroline's masters, such as James, Joyce, and many others.

How to Read a Novel is also a defense of the kind of fiction Caroline herself wrote, one that demands much of the reader. She states, "No matter how intelligent we are, as readers we are always more interested in what a character in a novel does than in what he thinks. All great novelists know this instinctively."[89] Many of her readers might dispute this assertion, but this credo is one reason Caroline Gordon is a writer's writer, not a popular author. Her works are often beautiful examples of technical mastery, but the thoughts, the feelings, the wit, and the humor that enlivened her letters and her conversations are absent from her characters and her authorial voice. In some ways the very seriousness with which she regarded the art of fiction barred her from the serendipitous, impulsive plunges into the human heart that often make for great fiction.

Despite teaching courses at NYU, City College, and Columbia, she finished *How to Read a Novel* in the spring of 1957 and turned it over to the publisher. After Max Perkins's death, her longstanding resentment at what she perceived as Scribner's lack of advertising for her books caused her to try a different publisher. *The Malefactors* had been published by Harcourt Brace; for *How to Read a Novel,* she turned to Viking. Although Scribner's would publish *Old Red and Other Stories* in 1963, she took her last novel, *The Glory of Hera,* to Viking, and her *Collected Stories* (1981) would be published by Farrar Straus Giroux. Her sense of lacking a strong and sympathetic editor when she was becoming increasingly estranged from Allen also diminished her productivity. In addition, all these changes of publisher did not help her reputation since it made her works harder to order and caused some difficulties with later reprintings.

As her career unraveled, so did her marriage. Allen and Caroline spent the summers of 1957 and 1958 apart while he taught at the Harvard Summer School. In the fall of 1958, he went to Oxford for the academic year while she remained at the Red House in Princeton and continued to commute to various New York colleges which now included the New School. Although Allen asked her to return to Minneapolis for the fall of 1959, if she thought they could live amicably together, the marriage was essentially over.[90] Caroline, however, was unwilling to concede defeat publicly by getting a divorce, a step that would also be contrary to the teachings of the Church. Allen, however, forced her hand by initiating a divorce suit himself.

Matters came to a head in the spring of 1959 when Cal Lowell reported to Caroline that Allen had found someone else, the poet Isabella Gardner. On May 8, 1959, Caroline wrote to Allen in Oxford: "Father McCoy [her spiritual director] has assured me all along that our marriage is indissoluble but I did not know until the other day that you wanted to marry somebody else. . . . I am going to see him [Father McCoy] this week and ask him if there is any way in which I can stop contesting your suit for divorce and if I can find any way to do this I promise you I will do so at once."

Caroline filed a countersuit charging Allen with "cruel and inhuman treatment," one of the possible grounds in Minnesota at that time. She chose that charge as one that would reveal least about their personal lives. Specifically, she charged that he had refused to live with her and suggested that she get a divorce, that he had sometimes not allowed her to return to their home, presumably in Minneapolis, and that his occasional drunkenness "brought shame and embarrassment upon her." They were divorced on August 18, 1959. Under the terms of the settlement, Caroline was to receive $375 a month for alimony until 1971, when it would be reduced to $300; she would be the beneficiary of all his life insurance policies; and she would own 145 Ewing Street, the Red House.[91]

On August 27, a little over a week after the divorce, Allen married Isabella Gardner. He wrote to Caroline to inform her, but also states that he and Caroline will always remain united in a certain way; in the next world, Allen concludes, he hopes he will be worthy of meeting her.[92]

Eileen Simpson, John Berryman's former wife, remembers Caroline's despair after the divorce:

> The second divorce tore Caroline apart. In this black and bitter period, she would invite me to dinner and pour out her anguish at Allen's having left her. That I had gone through two agonizing years after separating from John, and knew what she must be feeling, she refused to believe. Our situations were different: John had not left me for another woman; I had left him. Only because he drove me to it, I reminded her. Brushing aside what seemed to her a quibble, she said that I had the satisfaction of knowing that John suffered greatly; whereas, from all reports, Allen was shamelessly happy. Above all, John had not remarried. Allen had. Any word of comfort I offered Caroline was certain to be the wrong one. There was little I could do but listen. She was inconsolable.[93]

At sixty-five years of age, after thirty-four years of marriage, an angry, bitter, and humiliated Caroline Gordon had to decide how best to spend the rest of her life alone.

CHAPTER 7

During the 1960s and early 1970s, Caroline made her home at the Red House. In addition to all her friends in Princeton and the New York metropolitan area, she could be close to the Woods on Hodge Road, now a family of six with the birth of Amelia or "Amy," in 1957. In the mid-1960s, she added a modern wing to the Red House. Although the wing meant more convenience, it also made the Red House a bit too large for a single person and income. She solved the problem by various means, such as taking in borders and dividing the house into a duplex. In many ways, the most satisfactory arrangement, from 1965 to 1969, was sharing the house with Cary Peebles, an editor at Rutgers University Press. She also shared Caroline's interest in the classics, and they travelled together to Greece in 1967, along with young Caroline Wood, in preparation for *The Glory of Hera*.

Caroline was still her energetic, ambitious self during this decade, but she found it increasingly difficult to focus her energies efficiently or constructively. Occasionally, she would drink too much, and broke three ribs in the summer of 1968.[1] She could hurt others as well when she had one too many; in her bitterness she would unleash her sharp tongue on whomever was around at the time.

This underlying sense of injustice also occasionally surfaced in her public persona, before a class or a lecture audience. William Tillson, whom she met at Purdue, was now on the faculty at Wilmington College in Wilmington, Ohio. He invited her to lecture there in March 1965. She enjoyed seeing people she had known in

Wilmington during the year she spent there as an adolescent, but both Tillson's and the local newspaper's accounts of the lecture indicate a hostility toward the world as she found it. Her lecture sounded dogmatic, full of statements like "I am appalled whenever I hear of the writing of any living author being studied on any campus" and "I came very near to drowning in the tide of my students' illiteracy."[2] Tillson recalled: "She was lashing out. . . . I never saw her quite so negative, I guess you would call it, although she spoke well and wittily."[3]

Although she sometimes had trouble staying in control, she was by no means out of control and taught with some success at the University of California at Davis in the academic year 1962–63, Purdue the following year, and Emory in the spring of 1965. Celeste Turner Wright of Davis remembered her as "a supportive friend and a charming, lovable, warm-hearted woman, though her tongue could be sharp."[4] William Stuckey of Purdue recalled, "One of the things that impressed me about Caroline during the months she was in Lafayette was her enthusiasm for ordinary experience. She was always game for a party, a drive to Turkey Run, where we went for picnics."[5] Official Purdue also appreciated Caroline, awarding her an honorary degree in 1967.

Malcolm Cowley spent two months at Davis during the fall Caroline taught there. She had been angry with him, on and off, ever since the first divorce in 1946 because she jumped to the conclusion that Cowley would naturally take Allen's part. Cowley never felt that way and wanted to remain friends with both Allen and Caroline. In letters before Cowley's arrival at Davis and during conversations there, they made it up and had many enjoyable times together.[6] Although this estrangement had a happy ending, Caroline's sense that people were somehow siding with Allen against her caused her to alienate many of her former friends in the literary world, isolating herself from those who could help both her work and her reputation.

At Emory, she maintained her friendship with a community of Trappist monks in nearby Conyers, Georgia.[7] As a respite from urban Atlanta, she was occasionally allowed to stay in the their guesthouse, on a lake complete with swans.[8] Sadly, her dearest friend in Georgia, Flannery O'Connor, had died on August 3,

1964, but Caroline had been able to visit her in the hospital two months before her death.[9]

Her published essays of this period were mainly studies of Ford Madox Ford and Flannery O'Connor, which reiterate points she had made about them in earlier pieces.[10] In 1971 Cooper Square Publishers reprinted hardbound editions of those of her books for which Scribner's held the copyright. Her two larger projects early in the decade are also essentially reworkings. The revised edition of *The House of Fiction* appeared in 1960. Another collection of her short fiction, *Old Red and Other Stories,* was published in 1963 by Scribner's.

To *Old Red and Other Stories,* she added "Emmanuele! Emmanuele!" and omitted from it eight stories which appeared in *The Forest of the South* (1945).[11] The order of the remaining stories is reversed; instead of moving from the past forward, as did the earlier collection, the stories range from the near past to the distant past, ending with the pioneer story "The Captive." The only exception is "One Against Thebes," which was her recent reworking of her first story, "Summer Dust," in the context of the myth of Heracles.[12] By assigning first place to "One Against Thebes," Caroline may be suggesting that her readers reappraise her work from her new Catholic and Jungian perspective, as she herself had done.

None of her remaining friends were able to provide the kind of editorial direction and encouragement Caroline had received from Ford Madox Ford, Maxwell Perkins, and Allen Tate, not through any fault of theirs, but because of Caroline's feeling that there was no one left of her earlier mentors' stature. She particularly felt the lack of Allen's advice since he would help her focus on one aspect of her subject when her ambitions led her to topics and plots too large for a single work. She had not learned to make those kinds of judgments herself and tended to get so carried away that she could not complete her projects.

After Caroline finished *The Malefactors* in 1955, she planned a new book, *A Narrow Heart: The Portrait of a Woman.* The work's precise nature shifted over the twenty years she spent on it, but its core was inspired by her Jungian analysis in Rome and her conversion to Catholicism. She wanted to write her autobiography and show how her life fit into a mythic and religious context shared by everyone, truly a monumental task.

Caroline never got beyond age nine in her account of herself, which may indicate that she found a public self-examination too painful. Instead she branched out into accounts of various ancestors, immediate and distant, in what became an attempt to dissect the course of Western civilization from a Catholic and Jungian point of view. Parts of this work appeared in journals, including two sections of her autobiography. Another portion, "A Walk with the Accuser," which was first published in *Transatlantic Review* in 1969, is a good example of the grandiosity of her plan.[13] The story is a fictional treatment of John Calvin's damnation or Caroline's condemnation of Protestantism from a Catholic point of view.

Basically, *A Narrow Heart* grew into two other book-length works, *The Glory of Hera,* published by Doubleday in 1972, and *The Joy of the Mountains,* an incomplete novel.[14] *The Glory of Hera* is a retelling of the story of Hera and her namesake, the hero Heracles. In an autobiographical sense, the unfaithful yet powerful and intelligent Zeus and the jealous yet strong Hera appear as another version of the Tates' marriage. The true theme of the work, however, is the way Heracles' battles with evil, particularly the serpents he strangled, prefigured Christ's triumph over the serpent Satan. The novel contains passages of bravura description and dialogue—indeed, some of Caroline's best writing—but it is a digressive work that could have been improved by editing, mainly cutting, as Caroline herself realized after publication.[15]

The other book-length work, *The Joy of the Mountains,* is a novel concerning Caroline's distant ancestor, the explorer Meriwether Lewis of the Lewis and Clark expedition. Caroline believed Thomas Jefferson betrayed his neighbor and kinsman, Meriwether Lewis, by publicly proclaiming his somewhat mysterious death a suicide brought on by inherited melancholia. In the last chapter Caroline intended to reveal that Lewis was murdered, and by whom, but she did not write that last chapter. When one considers the quality of the manuscript, much more tightly controlled than *The Glory of Hera,* this seems quite a loss. An excerpt from *The Joy of the Mountains,* "The Strangest Day in the Life of Captain Meriwether Lewis," was published in the *Southern Review* in 1976.

Two other unrealized publications were a collection of the lectures she had delivered at St. Mary's College in the spring of 1964 and another book, *Creative Writing or Craft Ebbing?*. In a letter written in 1971, she explained the latter work's thesis: "The craft, I believe, is ebbing fast, partly as the result of the emphasis on creativity as opposed to technique."[16] Part of the problem with these two abortive works was simply the times. The exuberantly liberal 1960s were not a decade receptive to Caroline's Catholic and conservative views on the values of discipline and a classical education.

Despite the temper of the times, Caroline still managed to find an appreciative audience through a curious chain of events. She had been intermittently in touch with Allen through letters as they discussed the schedule of alimony payments, sometimes increasing them for her need or decreasing them for his. By 1971 Allen was the father of two sons by his third wife, the former Helen Heinz of St. Paul. According to the terms of his divorce from Caroline, he could decrease his payments from $375 to $300 in 1971. Because of his increasing responsibilities, he wished to do so.

As a result of their ensuing dispute, Caroline wanted to start earning money again, despite the willing aid of her son-in-law, Dr. Percy Wood. She was invited to give a series of lectures at the University of Dallas, a Catholic college in Irving, Texas, during their interterm early in 1973. She fell in love with the place and the people. They appreciated her conservative, Catholic views on education, and she needed them to maintain her pride and independence. Caroline was particularly struck by the President of the University and his wife, Donald and Louise Cowan, whom she termed "geniuses." Of Louise Cowan, who had earlier published a seminal book on the Fugitive poets, Caroline wrote, "She is the only woman I have ever known with a first-rate critical mind."[17] At the age of seventy-eight, Caroline decided to move to Texas and teach at the University of Dallas.

The first two years went fairly well as she tried to establish a creative writing program that reflected her views of a basic education. She wanted students to know some mythology, theology, Latin and Greek, logic, history, grammar, and the great books before they even attempted to write anything themselves. As she had in her earlier teaching, she devoted much time and attention to individual conferences and written commentaries on student work.

She reveled in it, writing, "I do love it out here, though. I simply adore it!"[18] Her happiness was augmented by the fact that her granddaughter and namesake Caroline Wood enrolled as a student there, and married a fellow Catholic, Christopher Fallon.

By 1975, when she was eighty years old, her health began to deteriorate, as did her ability to cope with the demands of teaching. She developed diverticulitis, and other ills, major and minor, which necessitated an often confusing round of medications. Because she was too frail to be left alone, she would have a friend or graduate student staying with her in her apartment on Northgate Drive. Although she did take leave in the spring and summer of 1975, she did not retire from teaching until December 1977 when she was eighty-two years old. The University of Dallas, which had already granted her an honorary degree in 1976, further showed their appreciation by a symposium honoring her in 1977. Old friends, such as Ashley Brown, Radcliffe Squires, and Howard Baker read papers in her honor, as did William Stuckey who recalled, "There was a party later on at the Cowans. The food and talk were wonderful, but it was all too much for Caroline. At several points she became confused and misunderstood what had been said and even got very angry. It was sad."[19]

Nancy and Percy Wood came to the rescue with an offer to share what was to be their retirement home in southern Mexico. Several years before, they had purchased a house at 7 Calle Comitan in the mountain city of San Cristobal de las Casas, part of the "Indian state" of Chiapas. Caroline had visited San Cristobal early in 1972 while the Woods were restoring their house and had a "blissful time." Through the Woods, Caroline purchased the house next door which was restored to adjoin the Woods' courtyard. Nancy installed Caroline there in the spring of 1978, with Caroline and Christopher Fallon to watch over her, along with her young great-grandson, Toby Fallon.

San Cristobal de las Casas is a remarkably beautiful city, both for its natural setting in the mountains and its lovely Spanish colonial architecture. The Woods' house on Calle Comitan has a panoramic view of the roofs and churches of the city and the mountains in the distance, as well as their own lushly flowering gardens on terraces below the house. Caroline, however, was too old to adapt to a new setting and new languages, Spanish and the local

Indian dialect. San Cristobal's remoteness provided further difficulty. To reach the city one must take a flight from Mexico City to
Tuxla Gutierrez in the lowlands and then journey by car for several
hours up the precipitous winding road to San Cristobal. Consequently, Caroline felt cut off from her old friends and literary
acquaintances, despite the city's active and interesting English-
speaking community. Regardless of this lengthy journey, Sally Fitzgerald and some of the friends Caroline shared with the Woods did
see come to see her.

Although she was no longer able to work, aside from sorting her
papers a little, Caroline was trying to assess her situation, and did
so with her usual self-mocking irony in a fragment entitled "A
Fixed Abode":

> All my life, I have craved, more than any other worldly good, a fixed
> abode. Divine Providence saw fit to make me a wanderer over the
> earth. I cannot recall, offhand, the names of all the cities and towns I
> have lived in.
>
> Old age came upon me like a clap of thunder. One day I was pursu-
> ing my ordinary avocations, with strength to spare for my neighbors'
> struggles. The next day I was a prisoner in an alien land. An arrhthymic
> heart which does not easily tolerate airplane flights, coupled with my
> family's fears for my safety, brought about this transformation.
>
> Once again, Providence has indulged in irony—perhaps its most ex-
> quisite as far as I am concerned. My abode is fixed, among mountains
> that tower higher than the cumulus clouds that sometime rest on their
> summits.[20]

Her abode was further fixed by a series of strokes that resulted in
circulatory problems and a gangrenous leg. Shortly after the leg
was amputated, Caroline's remaining strength ebbed. On April 11,
1981, at the age of eighty-six, Caroline Gordon died, with one
hand holding that of her daughter Nancy and the other holding the
hand of a priest, Jan de Voos, on whose face her eyes remained
focused. Perhaps through this man she saw that fixed abode toward which her life was directed.

APPENDIX: A Simplified Chart of Caroline Gordon's Ancestors

Paternal: The Gordons
William Fitzhugh Gordon
  soldier-statesman
William Fitzhugh Gordon
  lawyer-poet
  m. Nancy Morris
James Maury Morris Gordon
  teacher-sportsman
  m. Nancy Minor Meriwiether
Morris, *Caroline Ferguson,*
  William Fitzhugh

Maternal: The Meriwethers
Nicholas "the Welshman"
  founder of the American branch
Nicholas, classmate of Jefferson
  m. Margaret Douglas
Dr. Charles Meriwether
  founder of Kentucky branch
  m. Nance Minor (second wife of
  three)
Charles Nicholas Meriwether of
  Woodstock
  three children

Capt. "Ned"
  Meriwether,
  killed at
    Sacramento

Caroline Douglas
  Meriwether
    Goodlett,
    founder of U.D.C.

Nancy Minor Meriwether
  m. John Dickens Ferguson
Caroline Champlain Ferguson
  m. Douglas Meriwether
Nance Minor Meriwether
  m. James Maury Morris
    Gordon
Morris, *Caroline F.,* William F.

NOTES

Chapter 1

1. CG's unpublished memoirs, *A Narrow Heart,* and her notes for the memoirs, are among her papers in the Princeton University Library. Three portions of her childhood reminiscences appeared as "Always Summer," *Southern Review,* VII (1971), 430–46; "Cock-Crow," *Southern Review,* I (1965), 554–69; and "A Narrow Heart: The Portrait of a Woman," *Transatlantic Review,* III (1960), 7–19. CG's comments in this chapter and the information and legends about family history are taken from this published and unpublished material unless otherwise indicated.
2. *William Fitzhugh Gordon* (New York: Neale, 1909). The information about the Gordon family in this chapter is derived from this book, Gordon's notes, and family letters, all found among her papers in the Princeton University Library.
3. (Albany: Munsell's, 1982). The information about the Meriwethers in this chapter is derived from this book, from Gordon's notes in her papers in the Princeton University Library, from conversations with her in March 1981, and from conversations with her daughter Nancy Tate Wood in March 1981 and February 1985.
4. *Joy of the Mountains,* and notes for it, are among CG's papers in the Princeton University Library.
5. The information in this paragraph is derived from CG's papers and from Josephine M. Turner's *The Courageous Caroline: Founder of the UDC* (Montgomery, Alabama: Paragon Press, 1965).

Chapter 2

1. CG's unpublished memoirs, *A Narrow Heart,* and her notes for the memoirs are among her papers in the Princeton University Library.

Three portions of her childhood reminiscences appeared as "Always Summer," *Southern Review,* VII (1971), 430–46; "Cock-Crow," *Southern Review,* I (1965), 554–69; and "A Narrow Heart: The Portrait of a Woman," *Transatlantic Review,* III (1960), 7–19. The information about her childhood and relatives in this chapter is taken from this unpublished and published material unless otherwise indicated.

2. Mrs. Morris (Polly Ferguson) Gordon to VAM, February 5, 1985.
3. Conversations with CG, March 1981.
4. Danforth Ross to VAM, February 26, 1985.
5. *Ibid.*
6. The information on Louisa Meriwether is derived from CG's papers and from conversations with Caroline Gordon in March 1981 and with her daughter Nancy Tate Wood in March 1981 and February 1985.
7. Danforth Ross to VAM.
8. Interview with Andrew Lytle, August 15, 1982.
9. Mrs. Morris Gordon to VAM, February 5, 1985, and conversations with Nancy Tate Wood, February 1985.
10. Interview with Andrew Lytle.
11. *Collected Stories* (New York: Farrar Straus Giroux, 1981), 122.
12. Excerpted and sent to VAM by Mrs. Oscar (Ursula) Beach, County Historian, Clarksville, Tennesssee, November 27, 1984.
13. Mrs. Morris Gordon to VAM, February 5, 1985.
14. *Ibid.*
15. *Ibid.*
16. Interviews with Nancy Tate Wood, February 1985.
17. *Collected Stories,* 6.
18. (New York: Cooper Square, 1971), 97.
19. This and other family correspondence is in CG's papers in the Princeton University Library.
20. From Catherine Patterson Maccoy as excerpted and sent to VAM by Mrs. Oscar Beach.
21. Mrs. Morris Gordon to VAM, February 5, 1985.
22. Letter to "Miss Ella," February 9, 1902, in CG's papers in the Princeton University Library.
23. Conversations with Nancy Tate Wood, February 1985.
24. Conversations with CG, March 1981.
25. Conversations with CG and Nancy Tate Wood, March 1981.
26. Information on Alexander Campbell from *The New Encyclopedia Britannica,* 15th edition, Micropaedia, II.
27. Conversations with Nancy Tate Wood, February 1985.
28. Information on Gordon's University School is from Oscar Beach as conveyed to VAM by Mrs. Oscar Beach.

29. CG's notes for *A Narrow Heart.*
30. *Collected Stories,* 6.
31. Mrs. Morris Gordon to VAM, February 5, 1985.
32. Mrs. Morris Gordon to VAM, May 18, 1985.
33. Registrar's Office, Bethany College, Bethany, West Virginia.
34. Conversations with Nancy Tate Wood, February 1985.
35. "How I Learned to Write Novels," *Books on Trial,* XV (1956), 112, 160.
36. 1913 *Bethany College Bulletin.*
37. "The Value of Greek," *The Bethany Collegian,* XXI (November 1912).
38. "How I Learned to Write Novels," 162.
39. *Ibid,* 161.
40. *Ibid,* 160.
41. Conversations with CG, March 1981.
42. *Ibid.*
43. Conversations with Nancy Tate Wood, February 1985.
44. Mrs. C. H. Moore to Mrs. Oscar Beach to VAM, November 27, 1984.
45. Mrs. James L. Major to VAM, November 27, 1984.
46. Mrs. C. H. Moore to VAM, November 27, 1984.
47. Mrs. James L. Major to VAM, November 27, 1984.
48. *Ibid.*
49. Mrs. Morris Gordon to VAM, February 5, 1985.
50. Conversations with Nancy Tate Wood, February 1985.
51. Conversations with CG, March 1981.
52. *Caroline Gordon as Novelist and Woman of Letters* (Baton Rouge: Louisiana State University Press, 1984), 3.
53. (Baton Rouge: Louisiana State University Press, 1959), 98.
54. Conversations with CG, March 1981.

Chapter 3

1. "Emblems" (1931) in *Collected Poems 1919–1976* (New York: Farrar Straus Giroux, 1977), 36.
2. Allen Tate, *Memoirs and Opinion, 1926–1974* (Chicago: Swallow Press, 1975), 6–8.
3. For the facts of Tate's background and early life, I am indebted to Radcliffe Squires, *Allen Tate: A Literary Biography* (New York: Pegasus, 1971).
4. *Memoirs and Opinions,* 20.
5. *Ibid,* 21.

6. *Ibid*, 10.
7. *Ibid*, 20.
8. *Ibid*.
9. *Ibid*, 7.
10. *Ibid*, 16–17.
11. "Feathertop: A Moralized Legend" in *The Complete Novels and Selected Tales of Nathaniel Hawthorne*, ed. Norman Holmes Pearson (New York: Modern Library, 1937), 1092–1106.
12. *Memoirs and Opinions*, 22–23.
13. *Allen Tate: A Literary Biography*, 19.
14. Donald Davidson, "Introduction," *The Fugitive: April, 1922–December, 1925* (Gloucester, Mass.: Peter Smith, 1967), iii–v.
15. Squires, *Allen Tate: A Literary Biography*, 23–26.
16. "Foreword," *The Fugitive*, (April 1923), [1].
17. "Caveat Emptor," *The Fugitive*, I (June 1922), [3].
18. *Ibid*, 35.
19. *Memoirs and Opinions*, 29.
20. *Allen Tate and the Augustinian Imagination: A Study of the Poetry* (Baton Rouge: Louisiana State University, 1983), 17.
21. "In Amicitia" in *Allen Tate and His Work: Critical Evaluations*, ed. Radcliffe Squires (Minneapolis: University of Minnesota Press, 1972), 15.
22. *Memoirs and Opinions*, 39–40.
23. The facts of the exchange are taken from Thomas Daniel Young's *Gentleman in a Dustcoat: A Biography of John Crowe Ransom* (Baton Rouge: Louisiana State University Press, 1976), 152–55.
24. *Memoirs and Opinions*, 41.
25. *Gentleman in a Dustcoat*, 154.
26. *Memoirs and Opinions*, 31–32.
27. *Ibid*, 32.
28. *Ibid*, 30–31.
29. "Two Winters with Allen Tate and Hart Crane" in *Allen Tate and His Work*, 27.
30. Quoted by Cowley in "Two Winters with Allen Tate and Hart Crane," 26.
31. *Ibid*.
32. AT to John Wheelwright, July 13 [1924], Brown University Library.
33. Squires, *Allen Tate: A Literary Biography*, 57.
34. *The Poetry Reviews of Allen Tate, 1924–1944*, ed. Ashley Brown and Frances Neel Cheney (Baton Rouge: Louisiana State University Press, 1983).
35. Squires, *Allen Tate*, 56; Malcolm Cowley to Lewis Simpson, August 25, 1968, Newberry Library; Cowley, "Two Winters with Allen Tate

and Hart Crane," 28. Subsequent notes also refer to Cowley's correspondence with AT and CG in the Newberry Library.

36. September 5, 1959.
37. *The Southern Mandarins: Letters of Caroline Gordon to Sally Wood, 1924–1937*, ed. Sally Wood (Baton Rouge: Louisiana State University Press, 1984), 12–14.
38. Interview with Nancy Tate Wood, February 5–7, 1985; *The Southern Mandarins*, 13; Letters from Caroline's parents in the Princeton University Library.
39. Marriage License Bureau, Office of the City Clerk, the City of New York, License Number 15691-25.
40. Interview with Nancy Tate Wood.
41. Laura Riding to the *New York Review of Books* (December 22, 1983).
42. Quoted in Squires, *Allen Tate: A Literary Biography*, 51.
43. Susan Jenkins Brown, *Robber Rocks: Letters and Memories of Hart Crane* (Middletown: Wesleyan University Press, 1970), 40.
44. Interview with Malcolm Cowley, October 28, 1984.
45. III, 5–6.
46. Malcolm Cowley to Mrs. Anna B. Davis, Personal Service Fund in the Newberry Library.
47. "A Fixed Abode," Fragment of *A Narrow Heart*, Princeton University Library.
48. *The Southern Mandarins*, 16.
49. Riding to *New York Review of Books*.
50. "A Fixed Abode."
51. "A Fixed Abode" and Interview with Nancy Tate Wood.
52. Undated, *The Southern Mandarins*, 17–18.
53. "A Fixed Abode."
54. *The Southern Mandarins*, 17.
55. Susan Jenkins Brown, *Robber Rocks*, 30.
56. *The Letters of Hart Crane*, ed. Brom Weber (New York: Hermitage House, 1952), 212.
57. *Robber Rocks*, 31.
58. *Ibid*, 40–41.
59. Interview, October 28, 1985.
60. *The Letters of Hart Crane*, 246.
61. *Allen Tate: The Man and His Work*, 32–33.
62. *Robber Rocks*, 14.
63. Review of *Robber Rocks, Southern Review*, VI (1970), 482.
64. *The Letters of Hart Crane*, 226.
65. *Ibid*.
66. *Ibid*, January 7, 1926, 232.
67. Cowley, "Two Winters with Allen Tate and Hart Crane," 31.

68. To Gorham Munson, April 5, 1926, *The Letters of Hart Crane,* 244.
69. Hart Crane to his mother, April 18, 1926, *The Letters of Hart Crane,* 245–46.
70. *Ibid.*
71. *Robber Rocks,* 56.
72. To his mother, April 18, 1926, *The Letters of Hart Crane,* 248.
73. May 15, 1926, *The Southern Mandarins,* 21–22.
74. March 23, 1926.
75. Undated, Princeton University Library.
76. *The Southern Mandarins,* 30.
77. Princeton University Library.
78. September 26, 1926, *The Southern Mandarins,* 27.
79. Undated, *The Southern Mandarins,* 29–30.
80. *Ibid.*
81. *Ibid,* 28, 30.
82. *Collected Poems,* 20–23.
83. Interview with Caroline Gordon, March 1981.
84. Crane to his mother, April 18, 1926, *The Letters of Hart Crane,* 247.
85. CG to Sally Wood, May 15, 1926, *The Southern Mandarins,* 22.
86. Undated, *The Southern Mandarins,* 23–24.
87. *The Southern Mandarins,* 26–27.
88. Interview with Malcolm Cowley, October 28, 1985.
89. *Ibid.*
90. AT to Malcolm Cowley, January 10, 1962.
91. Interview with Malcolm Cowley.
92. "Ford Madox Ford" in *The Presence of Ford Madox Ford: A Memorial Volume of Essays, Poems, and Memoirs,* ed. Sondra J. Stang (Philadelphia: University of Pennsylvania Press, 1981), 13.
93. *Ibid,* 17.
94. Untitled in *The Presence of Ford Madox Ford,* 200.
95. January 21, 1930, *The Southern Mandarins,* 51.
96. "The Story of Ford Madox Ford" in *Highlights of Modern Literature: A Permanent Collection of Memorable Essays,* ed. Francis Brown (New York: NAL, 1954), 114, 116.
97. *A Good Soldier: A Key to the Novels of Ford Madox Ford* (Davis: University of California Library, 1963), Chapbook No. 1, 2.
98. *A Good Soldier* and "The Story of Ford Madox Ford."
99. "The Story of Ford Madox Ford," 116.
100. Interview with Andrew Lytle, August 15, 1982.
101. "Allen Tate: Upon the Occasion of His Sixtieth Birthday" in *Allen Tate: The Man and His Work,* 24.
102. "Allen Tate: A Portrait" in *Allen Tate: The Man and His Work,* 22.

103. Quoted in Joan Givner's *Katherine Anne Porter: A Life* (New York: Simon and Schuster, 1982), 178.
104. Interview with Malcolm Cowley; undated, *The Southern Mandarins*, 35.
105. Quoted in Givner, *Katherine Anne Porter*, 178.
106. Interview with Malcolm Cowley.
107. Undated, *The Southern Mandarins*, 35.
108. Interview with Caroline Gordon.
109. Undated, *The Southern Mandarins*, 35.
110. Undated, *The Southern Mandarins*, 39.
111. Princeton University Library.
112. Undated, *The Southern Mandarins*, 36.
113. CG to Virginia Tunstall, undated, University of Virginia Library, Charlottesville.
114. *Ibid.*
115. October 8, 1928, *The Southern Mandarins*, 41–42.
116. October 24, 1928, *The Literary Correspondence of Donald Davidson and Allen Tate,* ed. John Tyree Fain and Thomas Daniel Young (Athens: University of Georgia Press, 1974), 217–18.
117. University of Virginia, Charlottesville.
118. October 8, 1928, *The Southern Mandarins*, 42.
119. December 3, 1928, *The Southern Mandarins*, 44.
120. CG to Josephine Herbst, undated, Yale University Library. Subsequent notes also refer to CG's letters to Herbst in this library.
121. AT to Louise Bogan, February 10, 1929, Amherst College Library, Amherst, Massachusetts.
122. *The Poetry Reviews of Allen Tate, 1924–1944,* 43.
123. *Memoirs and Opinions,* 46–48.
124. CG to her mother, December 19, 1928, Princeton University Library.
125. December 3, 1928, *The Southern Mandarins*, 44.
126. December 19, 1928, Princeton University Library.
127. *Memoirs and Opinions,* 54.
128. CG to Josephine Herbst, February 11, 1929, Yale University Library.
129. *Memoirs and Opinions,* 48–49.
130. CG to Josephine Herbst, February 11, 1929.
131. *Memoirs and Opinions,* 131.
132. July 9, 1929, *The Southern Mandarins*, 48.
133. "Afterword" to *Aleck Maury, Sportsman* (Carbondale: Southern Illinois University Press, 1980), 294.
134. July 9, 1929, *The Southern Mandarins*, 45.
135. *Ibid,* 47.
136. AT to Raymond Holden, February 10, 1929, Amherst College Library, Amherst, Massachusetts.

137. February 11, 1929.
138. *Literary Correspondence of Donald Davidson and Allen Tate,* 223.
139. January 21, 1930, *The Southern Mandarins,* 52.
140. CG to Josephine Herbst.
141. *Memoirs and Opinions,* 50–51.
142. July 9, 1929, *The Southern Mandarins,* 46.
143. CG to Josephine Herbst, March 22, 1929.
144. Interview with Caroline Gordon, March 1981.
145. *Memoirs and Opinions,* 56–57.
146. October 6, 1929.
147. *Ibid.*
148. CG to Josephine Herbst, undated; *Memoirs and Opinions,* 57.
149. *Memoirs and Opinions,* 62.
150. *F. Scott Fitzgerald: A Biography* (Garden City, N.Y.: Doubleday, 1983), 235.
151. Interview with Caroline Gordon, March 1981.
152. *Memoirs and Opinions,* 60.
153. *Ernest Hemingway: Selected Letters, 1917–1961* (New York: Scribner's, 1981), 246.
154. *Memoirs and Opinions,* 63–64.
155. *Ernest Hemingway: Selected Letters,* 316.
156. *Literary Correspondence of Donald Davidson and Allen Tate,* 245.
157. Undated.
158. *Ernest Hemingway: Selected Letters,* 315.
159. *Memoirs and Opinions,* 47.
160. *Ibid,* 64.
161. "Afterword," *Aleck Maury, Sportsman,* 294.
162. *Ibid.*
163. *Memoirs and Opinions,* 54–57.
164. Interview with Caroline Gordon, March 1981.
165. *Memoirs and Opinions,* 54, 57.
166. Interview with Sally Wood, October 24, 1981.
167. Princeton University Library.
168. October 6, 1929.
169. February 4, 1930, Princeton University Library.
170. January 21, 1930, *The Southern Mandarins,* 51.

Chapter 4

1. CG to Sally Wood, January 21, 1930. *The Southern Mandarins: Letters of Caroline Gordon to Sally Wood, 1924–1937,* ed. Sally Wood (Baton Rouge: Louisiana State University Press, 1984), 50.

2. *The Forest of the South* (New York: Scribner's, 1945).
3. *Ibid.*
4. Maxwell Perkins to CG, March 13, 1930, Princeton University Library. Subsequent notes also refer to the Princeton University Library's correspondence between CG and Perkins.
5. Bernard Bandler to CG, January 21, 1930, Princeton University Library.
6. Ford Madox Ford to CG, February 24, 1930, Princeton University Library.
7. January 21, 1930, *The Southern Mandarins,* 50–51.
8. James Maury Gordon to CG, December 12, 1929, Princeton University Library.
9. July 31, 1930, *The Southern Mandarins,* 54.
10. *I'll Take My Stand* (Baton Rouge: Louisiana State University Press, 1977), xxxvii. Subsequent references to this edition will be indicated within the text.
11. AT to John Wheelwright, November 8, 1931, Brown University Library.
12. *Ibid.*
13. "In Amicitia" in *Allen Tate and His Work: Critical Evaluations* (Minneapolis: University of Minnesota Press, 1972), 16.
14. October [3?], 1930, *The Southern Mandarins,* 60.
15. *The Southern Mandarins,* 68–69.
16. CG to Lincoln Kirstein, undated, Yale University Library.
17. CG to Josephine Herbst, undated, Yale University Library. Subsequent references to letters from CG to Herbst in the Yale University Library will be noted only when they are dated.
18. Danforth Ross to VAM, February 26, 1985.
19. Norma H. Struss to VAM, May 14, 1985.
20. July 31, 1930, *The Southern Mandarins,* 55.
21. August 20, 1931, *The Southern Mandarins,* 82.
22. *The Southern Mandarins,* 64.
23. Interview with VAM, October 28, 1984.
24. October 20, 1930, *The Southern Mandarins,* 61–62.
25. CG to Herbst, undated.
26. February 21, 1931, *The Southern Mandarins,* 73.
27. *Ibid.*
28. Edmund Wilson to AT, August 16, 1931, *Letters on Literature and Politics, 1912–1972,* ed. Elena Wilson (New York: Farrar, Straus and Giroux, 1977), 214.
29. Edmund Wilson, "The Tennessee Agrarians," *New Republic,* LXII (1931), 279–81.

30. August 20, 1931, *The Southern Mandarins*, 82–83.
31. *The Forest of the South.*
32. *Ibid.*
33. My analysis of this story relies heavily on W. J. Stuckey, *Caroline Gordon* (New York: Twayne, 1972), 124.
34. CG to Perkins, January 21, 1930, Princeton University Library.
35. Correspondence between CG and Perkins in the Princeton University Library.
36. CG to Perkins, October 4, 1930, and Perkins to CG, November 14, 1930.
37. October 20, 1930, *The Southern Mandarins*, 62.
38. CG to Perkins, undated; Perkins to CG, March 20, 1931.
39. *The Southern Mandarins*, 78.
40. August 21, 1931, *The Southern Mandarins*, 86.
41. Danforth Ross to VAM.
42. Norma H. Struss to VAM.
43. November 2, 1931, *The Southern Mandarins*, 91.
44. CG to Josephine Herbst, undated.
45. Robert Penn Warren, "The Fiction of Caroline Gordon," *Southwest Review*, XX (1935), 6.
46. CG to Herbst, undated; Stark Young to CG, December 18, 1931, *Stark Young: A Life in the Arts: Letters, 1900–1962,* ed. John Pilkington (Baton Rouge: Louisiana State University Press, 1975), 382–83.
47. CG to Janet Lewis, November 12, 1931, Stanford University Libraries; CG to Sally Wood, undated, *The Southern Mandarins*, 87. Subsequent notes also refer to CG and AT's letters to Janet Lewis in the Stanford University Libraries.
48. October 31, 1931, University of Virginia Library, Charlottesville. Subsequent notes also refer to AT and CG's letters to Tunstall in this library.
49. November 2, 1931, *The Southern Mandarins*, 91.
50. November 18, 1931, *The Southern Mandarins*, 93.
51. August 1, 1931.
52. May 4, 1931, *The Southern Mandarins*, 76.
53. Undated.
54. New York: Torch Press, 1910.
55. CG to Perkins, undated.
56. Catherine B. Baum and Floyd C. Watkins, eds. "Caroline Gordon and 'The Captive': An Introduction," *Southern Review,* VII (1971), 460–61.
57. Perkins to CG, January 18, 1932.

58. CG to Robert Penn Warren, undated, Yale University Library. Subsequent notes also refer to CG's letters to Warren and his first wife Cinina in this library.

59. January 18, 1932.

60. CG to Robert Penn Warren, undated.

61. AT to John Wheelwright, January 21, 1932, Brown University Library; CG to Cinina Warren, undated.

62. Conversations with Nancy Tate Wood, April 1981.

63. Undated, and April 14, 1932, *The Southern Mandarins*, 99, 105.

64. AT to Janet Lewis, April 21, 1932.

65. Undated, *The Southern Mandarins*, 112.

66. May 3, 1932, *The Southern Mandarins*, 109.

67. *Ibid*, 114–116.

68. Information about pre-voyage travels from a letter from CG to Robert Penn and Cinina Warren, postmarked August 6, 1932.

69. May 4, 1931, *The Southern Mandarins*, 110.

70. CG to Malcolm and Muriel Cowley, August 4, [1932], Newberry Library, Chicago. Subsequent notes also refer to CG and AT's letters to Cowley in this library.

71. CG to Robert Penn and Cinina Warren, postmarked August 6, 1932.

72. August 4, [1932].

73. Thomas H. Landess, "Caroline Gordon's Ontological Stories," in *The Short Fiction of Caroline Gordon*, ed. Landess (Irving, Texas: University of Dallas Press, 1977), 58.

74. February 10, 1932, *The Southern Mandarins*, 136–37.

75. Undated, *The Southern Mandarins*, 99, 100.

76. Interview with Sally Wood, October 24, 1981.

77. December 10, 1932, *The Literary Correspondence of Donald Davidson and Allen Tate*, ed. John Tyree Fain and Thomas Daniel Young (Athens: University of Georgia Press, 1974), 280.

78. *Collected Poems, 1919–1976* (New York: Farrar Straus Giroux, 1977), 66–67.

79. Undated and November 23, 1932, *The Southern Mandarins*, 122.

80. *The Southern Mandarins*, 124.

81. Undated.

82. Undated, *The Southern Mandarins*, 135.

83. December 1, 1932, *The Southern Mandarins*, 126–27.

84. Undated, *The Southern Mandarins*, 131.

85. February 10, 1933, *The Southern Mandarins*, 137.

86. June 15, 1932, *The Southern Mandarins*, 115.

87. *The Forest of the South*.

88. Undated, *The Southern Mandarins*, 132.
89. Allen Tate, *Memoirs and Opinions, 1926–1974* (Chicago: Swallow Press, 1975), 67.
90. Undated, *The Southern Mandarins*, 139–40.
91. Interview with Nancy Tate Wood, February 5–7, 1985.
92. "The Meriwether Connection" in *The Dream of the Golden Mountains: Remembering the 1930s* (New York: The Viking Press, 1980), 194.
93. Interview with Malcolm Cowley, October 28, 1984.
94. August 21, 1931, *The Southern Mandarins*, 85.
95. June 15, 1932, *The Southern Mandarins*, 116.
96. "The Meriwether Connection," 197–98. In this and the next quotation from this article, Cowley does not identify the woman as Marion Henry; I do.
97. *Ibid.,* 199.
98. Interview with Malcolm Cowley.
99. June 6, 1933, *The Southern Mandarins*, 143–44.
100. November 27, 1934, *The Southern Mandarins*, 172.
101. James Maury Gordon to CG, undated, Princeton University Library.
102. "Afterword," *Aleck Maury, Sportsman* (Carbondale: Southern Illinois University Press, 1980), 295.
103. CG to Perkins, undated.
104. August 14, 1933, *The Southern Mandarins*, 147.
105. Undated, *The Southern Mandarins*, 151.
106. October 1933–January 1934, *The Southern Mandarins*, 153–60.
107. May 22, 1934, *The Southern Mandarins*, 165.
108. October 1, 1934, *The Southern Mandarins*, 166–68.
109. Undated, *The Southern Mandarins*, 169.
110. *The Forest of the South.*
111. *Ibid.*
112. CG to Robert Penn Warren, undated.
113. February 5, 1935, *The Southern Mandarins*, 175.
114. CG to Robert Penn Warren, undated.
115. June 15, 1935, *The Southern Mandarins*, 188.
116. CG to Mark Van Doren, undated, Columbia University Library. Subsequent notes also refer to CG's letters to Van Doren in this library.
117. November 27, 1934, *The Southern Mandarins*, 173.
118. Undated, Columbia University Library.
119. Undated.
120. October 1, 1934 and February 15, 1935, *The Southern Mandarins*, 166, 176.

121. February 15, 1935, *The Southern Mandarins,* 175.

122. Undated, *The Southern Mandarins,* 183–84.

123. All information on this conference is from Thomas W. Cutrer's "Conference on Literature and Reading in the South and Southwest, 1935," *Southern Review,* XXI (Spring 1985), 260–300.

124. Undated, *The Southern Mandarins,* 185.

125. March 6 and March 23, 1935, *The Southern Mandarins,* 178, 180.

126. Danforth Ross to VAM.

127. March 23, 1935, *The Southern Mandarins,* 181.

128. Program Note to the Minnesota Production of *The Governess,* University of Minnesota Library.

129. February 15, 1935, *The Southern Mandarins,* 175.

130. Undated, *The Southern Mandarins,* 192–93.

131. August 8, 1935, *The Southern Mandarins,* 194.

132. October 27, 1935, *The Southern Mandarins,* 196–97.

133. CG to Robert Penn Warren, undated.

134. *The Forest of the South.*

135. "The Idea of Nature and the Sexual Role in Caroline Gordon's Early Stories of Love," in *The Short Fiction of Caroline Gordon: A Critical Symposium,* 105.

136. *Dorothy Day: A Biography* (San Francisco: Harper & Row, 1982), 287–88.

137. *The Long Loneliness* (San Francisco: Harper & Row, 1951, 1981), 171.

138. AT to Donald Davidson, *The Literary Correspondence of Donald Davidson and Allen Tate,* 297.

139. Interview with Nancy Tate Wood, February 5–7, 1985.

140. September 10, 1936, *The Southern Mandarins,* 201.

141. *Ibid.*

142. *Ibid,* 202.

143. Interview with Nancy Tate Wood.

144. Undated letters.

145. January 13, 1937, University of Virginia Library, Charlottesville.

146. *The Forest of the South.*

147. Many critics have noted Gordon's use of natural symbols in "The Brilliant Leaves." In particular, see Andrew Lytle's "Caroline Gordon and the Historic Image," *Sewanee Review,* LVII (1949), 562–67; Louise Cowan, "Nature and Grace in Caroline Gordon," *Critique* I (1956), 11–27; James E. Rocks, "The Short Fiction of Caroline Gordon," *Tulane Studies in English,* XVIII (1970), 115–35; and John E. Alvis, "The Idea and Nature of the Sexual Role in Caroline Gor-

don's Early Stories of Love" in *The Short Fiction of Caroline Gordon*, 85–111.

148. May 3, 1937, Yale University Library.
149. January 8, 1937, *The Southern Mandarins*, 202–205.
150. *None Shall Look Back* (New York: Scribner's, 1937), 357. Subsequent references to this novel will be indicated within the text.
151. Undated, *The Southern Mandarins*, 208.
152. Undated, *The Southern Mandarins*, 206.
153. *A Life in Letters*, 752–53.
154. "Dulce et Decorum Est," *New Republic* (March 31, 1937), 244–45.
155. Scribner Files, Princeton University Library.
156. *Ibid.*
157. "Visiting the Tates" in *Allen Tate and His Work*, 34.
158. Undated, *The Southern Mandarins*, 209.
159. "Visiting the Tates," 34.
160. Undated, *The Southern Mandarins*, 208.
161. July 10, 1937, *The Southern Mandarins*, 210.
162. Quoted in Ian Hamilton's *Robert Lowell: A Biography* (New York: Random House, 1982), 49.
163. *Ibid.*
164. July 10, 1937, *The Southern Mandarins*, 211.
165. *Ibid.*
166. Joan Givner, *Katherine Ann Porter: A Life* (New York: Simon and Schuster, 1982), 304.
167. October 16, 1937, *The Southern Mandarins*, 214.
168. *Katherine Ann Porter*, 305.

Chapter 5

1. CG to Maxwell Perkins, undated, Princeton University Library.
2. James Ross to VAM, October 29, 1984.
3. CG to Perkins, undated, Princeton University Library.
4. *Green Centuries* (New York: Scribner's, 1941), 458–59. Subsequent references to this novel will be indicated in the text.
5. *The Forest of the South* (New York: Scribner's, 1945).
6. *The Fathers and Other Fiction* (Baton Rouge: Louisiana State University Press, revised edition 1977), 5.
7. *The Autobiography of Mark Van Doren* (New York: Harcourt Brace, 1958), 258.
8. CG to Robert Lowell, undated, McFarlin Library, University of Tulsa, Tulsa, Oklahoma.

9. AT to Lowry Axley, February 19, 1939, Georgia Historical Society, Savannah, Georgia.

10. Information on the Tates at this conference is from Ben C. Toledano's "Savannah Writers' Conference—1939," *Georgia Review,* III (1968), 148–59; and the Georgia Historical Society.

11. AT to Donald Davidson, May 12, 1939, *The Literary Correspondence of Donald Davidson and Allen Tate,* ed. John Tyree Fain and Thomas Daniel Young (Athens: University of Georgia Press, 1974), 319–20.

12. AT to Paul Green, January 2, 1939 [1940], Southern Historical Collection, University of North Carolina, Chapel Hill.

13. *Allen Tate: A Literary Biography* (New York: Pegasus, 1971), 154–55.

14. CG to Muriel Cowley, undated, Newberry Library, Chicago. Subsequent references to CG and AT's letters to Malcolm and Muriel Cowley at the Newberry will be noted only if they are dated.

15. CG to "Little May" Morse, undated, Kentucky Library, Western Kentucky University, Bowling Green.

16. Malcolm Cowley's interview with VAM, October 28, 1984.

17. CG to "Little May" Morse, undated.

18. Bobbs-Merrill Mss., Lilly Library, Indiana University, Bloomington.

19. CG to "Little May" Morse, undated.

20. Russell Fraser, *A Mingled Yarn: The Life of R. P. Blackmur* (New York: Harcourt Brace Jovanovich, 1981), 244.

21. Undated, Bobbs-Merrill Mss.

22. Interviews with VAM, February 5–7, 1985.

23. CG to Ward Dorrance, undated, Southern Historical Collection, University of North Carolina, Chapel Hill. Subsequent references to CG's letters to Dorrance in this collection will only be noted when they are dated.

24. Postmarked November 8, 1941, Columbia University Library.

25. *A Mingled Yarn,* 194.

26. CG to "Little May" Morse, undated, Kentucky Library.

27. CG to Mark Van Doren, undated, Columbia University Library.

28. AT to Malcolm Cowley, October 5, 1942.

29. CG to Mark Van Doren, undated, Columbia University Library.

30. *The Women on the Porch* (New York: Scribner's, 1944), 185.

31. CG to Dorrance, undated.

32. CG to Malcolm Cowley, undated.

33. CG to Jean Stafford, undated, McFarlin Library, University of Tulsa, Tulsa, Oklahoma. Subsequent letters from CG to Stafford will be noted only if they are dated.

34. Ian Hamilton, *Robert Lowell: A Biography* (New York: Random House, 1982), 82.

35. AT to Virginia Tunstall, January 19, 1943, University of Virginia Library, Charlottesville.
36. *Collected Poems, 1919–1976* (New York: Farrar Straus Giroux, 1977), 145–61.
37. CG to Stafford, undated.
38. Interview with the Cheneys, August 1982.
39. *Ibid.*
40. CG to Jean Stafford, undated.
41. CG to Jean Stafford [November 19, 1943].
42. *Ibid.*
43. Information about her father's final illness from CG's undated letters to AT and the Cheneys, Princeton University Library.
44. CG to AT, undated letter, Princeton University Library.
45. Interview with the Cheneys, August 1982.
46. Conversations with Nancy Tate Wood, February 5–7, 1985. See also Joan Givner, *Katherine Ann Porter: A Life* (New York: Simon and Schuster, 1982), 330–31.
47. Conversations with CG, March 1981.
48. *The Forest of the South* (New York: Scribner's, 1945).
49. Andrew Lytle, "The Forest of the South," *Critique,* (1965), 6–7.
50. *The Forest of the South.*
51. M. E. Bradford, "The High Cost of 'Union': Caroline Gordon's Civil War Stories,' in *The Short Fiction of Caroline Gordon: A Critical Symposium,* ed. Thomas H. Landess (Irving,Texas: University of Dallas Press, 1972), 116.
52. *The Forest of the South* (1945).
53. John E. Alvis, "The Idea of Nature and the Sexual Role in Caroline Gordon's Early Stories of Love" in *The Short Fiction of Caroline Gordon,* 100.
54. Lytle, "The Forest of the South," 5.
55. Information about settling into the Robert E. Lee from an undated letter from CG to Stafford.
56. Information on CG's state of mind from undated letters to Stafford.
57. AT to Louise Bogan, April 10, 1945, Amherst College Library, Amherst, Massachusetts.
58. CG to Percy Wood, Princeton University Library.
59. Information on Loulie's final visit from CG's undated letters to Stafford.
60. *Collected Stories* (New York: Farrar Straus Giroux, 1981), 306–16.
61. AT to Louise Bogan, April 10, 1945, Amherst College Library, Amherst, Massachusetts.
62. CG to Jean Stafford, undated.
63. *Ibid.*
64. Interview with Nancy Tate Wood, February 5–7, 1985.

Chapter 6

1. CG to Jean Stafford, undated letters, McFarlin Library, University of Tulsa, Tulsa, Oklahoma. Subsequent references to letters from CG to Stafford will only be noted if the letters are dated.
2. Information on Stafford and Lowell in the fall of 1945 from Ian Hamilton's *Robert Lowell: A Biography* (New York: Random House, 1982), p. 102. Information on CG's plans from an undated letter to Malcolm Cowley, Newberry Library, Chicago. Subsequent references to CG's letters to Malcolm or Muriel Cowley in this collection will only be noted if they are dated.
3. Undated letter, Southern Historical Collection, University of North Carolina Library, Chapel Hill. Subsequent references to letters from CG to Ward Dorrance in this collection will be noted only if they are dated.
4. Undated letters to Jean Stafford and Malcolm Cowley.
5. November 2, 1945, Newberry Library.
6. CG to Malcolm Cowley.
7. CG to Dorrance.
8. CG to Dorrance.
9. CG to Dorrance.
10. CG to Dorrance.
11. CG to AT, undated. Princeton University Library. Subsequent references to correspondence between the Tates in this collection will be noted only if they are dated.
12. "Mr. Faulkner's Southern Saga: Revaluing his Fictional World and the Unity of Its Patterns," *New York Times Book Review* (May 5, 1946), 1, 45.
13. CG to Dorrance.
14. CG to Dorrance.
15. CG to AT.
16. CG to Margaret "Piedie" Campbell, undated, Kentucky Library, Western Kentucky University, Bowling Green, Kentucky.
17. CG to Dorrance.
18. CG to Dorrance.
19. *The Collected Stories of Caroline Gordon* (New York: Farrar Straus Giroux, 1981), 3–15.
20. Baptismal Certificate in the Caroline Gordon Papers, Princeton University Library.
21. "The Art and Mystery of Faith," *Newman Annual* of the University of Minnesota's Newman Foundation (December 1953), 55–62.
22. Flannery O'Connor to "A," May 19, 1956, *The Habit of Being: Letters*

of Flannery O'Connor, ed. Sally Fitzgerald (New York: Farrar Straus Giroux, 1979), 159.

23. *Collected Stories,* 105–20.

24. Quoted as the epigraph to *The House of Fiction: An Anthology of the Short Story with Commentary* by Caroline Gordon and Allen Tate (New York: Scribner's, 1950).

25. *The House of Fiction,* vii.

26. *The House of Fiction,* vii–viii.

27. "Notes on Faulkner and Flaubert," *Hudson Review* I (1948), 222–31. "Stephen Crane," *Accent,* IX (Spring 1949), 153–57. "Notes on Hemingway and Kafka," *Sewanee Review,* LVII (Spring 1949), 215–26. "Notes on Chekhov and Maugham," *Sewanee Review,* LVII (Summer 1949), 401–10.

28. W. J. Stuckey to VAM, January 17, 1985.

29. Information from CG to Ward Dorrance, undated, and AT to John Wheelwright of Scribner's, July 21, 1948, Princeton University Library.

30. CG to Nancy Tate Wood, undated, Princeton University Library. Subsequent references to letters from CG to Nancy Tate Wood in this collection will be noted only if they are dated.

31. CG to Dorrance.

32. Information on the Lowell visit from Hamilton's *Robert Lowell,* 155–56.

33. CG to Dorrance.

34. CG to Dorrance.

35. CG to Dorrance.

36. Eileen Simpson, *Poets in Their Youth: A Memoir* (New York: Random House, 1982), 197–98.

37. *The Strange Children* (New York: Scribner's, 1951), 107. Subsequent references to this novel will be indicated within the text.

38. Quotation and information about CG's intent from letter to Dorrance, "Octave Day of All Saints, 1950."

39. "The Art and Mystery of Faith," 58.

40. An undated letter to Flannery O'Connor quoted in "A Master Class: From the Correspondence of Flannery O'Connor," *Georgia Review,* XXXIII (Winter 1979), 834.

41. Postmarked February 27, 1951.

42. Interview with Malcolm Cowley, October 28, 1984.

43. Danforth Ross, "Caroline Gordon's Golden Ball," *Critique,* (1956), 69.

44. Robert Kettler to VAM, November 1984.

45. Interview with Jean Detre, October 26, 1984.

46. Ross, "Caroline Gordon's Golden Ball," 69.
47. Interview with Jean Detre.
48. Kettler to VAM.
49. *Ibid.*
50. Marjorie Kaplan to VAM, January 1985.
51. *Ibid.*
52. Correspondence of Wheelright, AT, and CG in early 1953, Scribner Collection, Princeton University Library.
53. January 8, 1953.
54. Fitzgerald, "A Master Class," 829.
55. "A Master Class," 831.
56. CG to Nancy Tate Wood, October 10, 1951.
57. AT to Donald Davidson, October 27, 1951, *The Literary Correspondence of Donald Davidson and Allen Tate,* ed. John Tyree Fain and Thomas Daniel Young (Athens: University of Georgia Press, 1974), 351
58. CG to Nancy Tate Wood.
59. *Ibid.*
60. Quoted in William D. Miller, *Dorothy Day: A Biography* (San Francisco: Harper & Row, 1982), 400.
61. Dated "Birthday of Our Lady."
62. CG to Malcolm Cowley.
63. "A Virginian in Prairie Country: Two New Studies Explore the Life and Work of Novelist Willa Cather," *New York Times Book Review* (March 8, 1953), 1, 31.
64. "Some Readings and Misreadings," *Sewanee Review,* LXI (1953), 384–407.
65. "Mr. Verver, Our National Hero," *Sewanee Review,* LXIII (1955), 29–47.
66. "The Art and Mystery of Faith," 56.
67. CG to Nancy Tate Wood.
68. *Collected Stories,* 335. Subsequent references to this story will be indicated within the text.
69. CG to Nancy Tate Wood.
70. CG to AT, May 26, 1953.
71. Allen Tate, *Collected Poems, 1919–1976* (New York: Farrar Straus Giroux, 1977), 138–39.
72. Information on summer travels and events in Rome to this point from letters of CG to Nancy Tate Wood.
73. CG to Stark Young, undated, Humanities Research Center, University of Texas, Austin.
74. CG to Nancy Tate Wood and Samuel Monk, December 8, 1953.

75. *Ibid.*
76. CG to Ward Dorrance, February 19, 1954.
77. *Ibid.*
78. Correspondence between AT and CG during May 1954.
79. CG to Stark Young, undated, Humanities Research Center University of Texas, Austin.
80. *The Malefactors* (New York: Harcourt Brace, 1956), 176. Subsequent references to this novel will be indicated within the text.
81. Based on the account in William Miller's *Dorothy Day,* 453–54.
82. CG to AT, December 5, 1955.
83. CG to Robert Lowell, undated and March 28, 1956, Houghton Library, Harvard University.
84. CG to AT, April 11, 1956.
85. CG to Nancy Tate Wood.
86. CG to Malcolm and Muriel Cowley.
87. Letters of AT to CG, July–September 1956.
88. CG to Malcolm Cowley, December 26, 1956.
89. *How to Read a Novel* (New York: Viking, 1957), 222.
90. AT to CG, August 13, 1958.
91. Records of the divorce at Hennepin County Court House, Minneapolis, Minnesota.
92. AT to CG, September 6, 1959.
93. Simpson, *Poets in Their Youth,* 199.

Chapter 7

1. CG to Nancy Tate Wood, July 2, 1968, Princeton University Library. The other letters to Nancy Tate Wood that are cited are also in this library.
2. Harry Daniels, "Writer Gordon Suggests Ban on Campus Writers," *Wilmington Monitor* (Friday, March 12, 1965), 6.
3. William Tillson to VAM, November 26, 1984.
4. Celeste Turner Wright to VAM, October 7, 1984.
5. William J. Stuckey to VAM, January 17, 1985.
6. Correspondence among Cowley, CG, and AT in the Newberry Library, Chicago; Cowley's interview with VAM, October 28, 1984.
7. To one of these monks, who wrote as "Jack English," CG wrote the "Letters to a Monk," *Ramparts,* III (1964), 4–10.
8. CG to William Tillson, April 24, 1965; CG to Nancy Tate Wood, January 10, 1965.
9. Robert Fitzgerald's Introduction to Flannery O'Connor's *Everything*

That Rises Must Converge (New York: Farrar Straus Giroux, 1965), vii–xxxiv.

10. "To Ford Madox Ford," *Transatlantic Review,* n.s. III (1960), 5–6. *A Good Soldier: A Key to the Novels of Ford Madox Ford,* Chapbook No. 1 (Davis: University of California Library, 1963). Untititled tribute to Flannery Ò'Connor, *Espirit,* VIII (Winter 1964) 28. "The Elephant" [Ford], *Sewanee Review,* LXXIV (1960), 856–71. "An American Girl" in *The Added Dimension: The Art and Mind of Flannery O'Connor,* ed. Melvin J. Friedman and Lewis A. Lawson (New York: Fordham University Press, 1966), 123–37. "Heresy in Dixie" *Sewanee Review,* LXXVI (1968), 263–97. Foreword to Sister Kathleen Feeley's *Flannery O'Connor: The Voice of the Peacock* (New Brunswick, N. J.: Rutgers University Press, 1972), ix–xii. "Rebels and Revolutionaries," *Flannery O'Connor Bulletin,* III (1974), 40–56.

11. "The Forest of the South," "The Ice House," "The Burning Eyes," "To Thy Chamber Window, Sweet," "The Long Day," "Mr. Powers," "Her Quaint Honour," and "The Enemies."

12. "One Against Thebes" appeared as "The Dragon's Teeth" in *Shenandoah,* XIII (1961), 22–34.

13. "A Walk With the Accuser" later appeared in the *Southern Review,* XIII (1977), 597–613.

14. CG's unfinished works, drafts, and notes are among her papers in the Princeton University Library.

15. CG to Robert Giroux, February 29, 1972, CG's file at Farrar Straus Giroux.

16. CG to Radcliffe Squires, May 4, 1971, Washington University Library, St. Louis, Missouri.

17. CG to Malcolm Cowley, July 5, 1973, Newberry Library.

18. CG to Eileen and Radcliffe Squires, February 19, 1974, Washington University Library, St. Louis, Missouri.

19. W. J. Stuckey to VAM.

20. CG's papers, Princeton University Library.

INDEX